Reading the Nervous System

A Physiological Reading of Autistic Experience

María E. Reyes, M.S., CCC-SLP

ISBN (Print): 979-8-9955583-0-9
First Edition
ISBN (eBook): 979-8-9955583-1-6
First Edition
Cover design by María E. Reyes
Interior design by María E. Reyes
Published by Neurodiversidad Consciente

York, PA, USA

Library of Congress Control Number: 2026912582

Permission requests and rights inquiries: ReadingTheNervousSystem.com

To my children, thank you for staying while I was learning how to stay.

To my mother, for giving what she could, when she could.

And to my life partner, for walking alongside me with presence, care, and steadiness.

TABLE OF CONTENTS

Acknowledgments

This book is grounded in years of clinical work alongside children and families in the early intervention community. What I know about how nervous systems organize experience under conditions of demand, recovery, and mismatch did not come from theory first. It came from watching bodies navigate environments not built around them, and from the families who stayed present through that navigation with care and steadiness. I am grateful to each of them.

The children I have accompanied over the years are present throughout this book in ways they will never see and I cannot fully name. Their bodies taught me what to look for before the words were available to describe it. This framework exists because they made visible what was otherwise invisible.

I am grateful to the autistic writers, researchers, educators, and community members whose published accounts are treated throughout this book as primary evidence of lived organization. Their willingness to articulate internal experience in public, with precision and without softening, made this framework more honest than it would otherwise have been.

Writing this required specific conditions I did not always create for myself. To my children, my mother, and my life partner: thank you for the presence that made recovery possible when I had used more than I had.

M.E.R.

Note from the Author

This book was built through questions and connections.

I did not begin with an answer I wanted to prove. I began with a need for coherence.

I wanted to understand how the nervous system organizes experience. I kept encountering dimensions that made sense when considered separately—sensory processing, interoception, prediction, relationship, context—but became more difficult to bring together when I tried to follow what was happening in a person across time. Each perspective explained something. I wanted to understand how the pieces belonged together.

Autism gave that search urgency.

Working with autistic children, I encountered questions I could not resolve by looking at behavior alone. Why could something be available in one moment and inaccessible in another? Why could a seemingly small demand produce a large response, while something more difficult might not? Why did time matter? Why did what happened earlier matter? Why could the same person appear so different under different sensory, relational, or environmental conditions?

The more closely I looked, the less interested I became in asking what a behavior meant in isolation. I wanted to understand the system organizing the conditions from which that behavior became possible.

So I kept asking questions.

What happens when we look at a response not in isolation, but in relation to what preceded it? What changes when we consider not only what is being asked of a system, but what it is already carrying? What becomes visible when we follow these relationships across time rather than treating each moment as a separate event?

Autism did not give me the answers. It gave me reasons to keep asking better questions.

I was not looking for a new explanation to replace the existing ones. I was trying to understand how the pieces could belong together.

Those questions led to other questions, and the connections between them gradually became the architecture developed in this book.

Reading the Nervous System is therefore not built from certainty. It is built from disciplined questions and theoretically grounded connections. Some of those connections are supported by established bodies of research; others are propositions that remain open to examination, refinement, and empirical testing.

The purpose is not to close interpretation around a person. It is to make our reasoning more careful: to distinguish what we observe from what we infer, to remain curious about what may be organizing the response we see, and to change our reading when new information asks us to.

In that sense, the questions are not simply how this book began. They are part of what the book is asking us to practice.

Introduction

It is early morning. A person enters a room where the day's work will begin. The lights hum faintly overhead. Voices move across the space. Before the mind has time to interpret anything, the body responds: breathing shifts slightly, muscles along the back organize into quiet readiness, attention scans the room and settles. Signals arrive from the environment and from within the body—temperature, pressure, movement, sound.

Some are filtered quickly and fade into the background. Others remain active, requiring continued management as the day's first tasks begin.

Most of this activity goes entirely unnoticed. Experience appears as thoughts, emotions, decisions, and actions. Yet beneath these layers, the nervous system organizes the conditions under which experience becomes possible. It regulates activation, integrates sensory input, allocates energy, and prepares the body for what comes next. Long before conscious interpretation or deliberate response, the body is at work. Experience does not begin with thought. It begins with organization.

This book follows a perspective emerging across neuroscience and philosophy of mind: experience arises from the body's ongoing regulation of its physiological state rather than from cognition operating on a neutral substrate. Converging work across interoception (Damasio, 2010; Craig, 2009), predictive processing (Friston, 2010; Seth, 2021), and enactive cognition (Varela et al., 1991; Thompson, 2007) points to a shared insight: what we experience as the world—and as ourselves within it—is shaped by the system's continuous effort to regulate itself in relation to its environment. Physiological organization does not develop apart from those conditions. It develops through continuous interaction between the organism and the environment within which it must sustain itself.

What these frameworks share is the recognition that regulatory activity is foundational. It is the process through which experience is constituted. What follows is a way of reading what becomes visible when this process is followed in lived time.

Within this framework, regulation is not understood as the absence of activation or emotional intensity. Living systems continuously shift, activate, and reorganize in relation to changing internal and external conditions. What matters physiologically is not the presence of activation alone, but how activation unfolds within the larger organization of the system over time.

What a person does emerges within the state of the system from which it becomes possible.

The cost of a new demand cannot be understood from the demand alone. It depends on the state in which that demand is encountered and on what has unfolded before it. What becomes visible are the four parameters through which any nervous system organizes experience across time: concurrent demand, activation slope, regulatory margin, and recovery.

These parameters describe all nervous systems. What differs across nervous systems is how these processes operate—and what those differences cost in environments built around specific assumptions.

When the assumptions of the environment repeatedly diverge from the conditions required by the system, regulation itself becomes more effortful and more visible across time.

Autistic experience provides a particularly clear view of these dynamics.

When system parameters differ significantly from environmental expectations—regarding sensory processing, recovery time, and pace of demand—regulatory processes become more visible. What appears as difficulty is physiology made visible under conditions of mismatch.

Autistic experience therefore serves this book in two ways. It is both the primary focus and the lens through which the underlying mechanics of nervous system organization become more legible.

From this perspective, autism is read at the level of nervous system configuration—how processing, accumulation, timing, and recovery organize across time—rather than first by diagnostic category or severity. This is the level at which the framework operates; it is not a claim about what autism is, but about what this book is reading.

This book does not propose a new theory of the nervous system or introduce new mechanisms. It draws on established principles across neuroscience, physiology, developmental science, and clinical observation, offering a way to read how these processes organize lived experience across time. Its aim is a reading practice: following nervous system organization so that behavior can be interpreted in relation to the physiological conditions within which it emerges.

This is a process-based framework describing movement—how states emerge, accumulate, and resolve—rather than fixed categories or traits. The parameters orient observation; they do not determine individual experience. Reading the nervous system is a gradual process unique to each organism and shaped by factors such as diagnosis, sensory organization, communication style, developmental history, and accumulated experience. This framework does not provide fixed interpretations or exact responses, but a way of observing what to look for as understanding develops through ongoing observation, adjustment, and relationship.

The scientific frameworks referenced—predictive processing, autonomic regulation, interoception, and allostatic load—are used instrumentally, each where its explanatory range is most relevant. Ongoing debates are addressed where necessary.

Terms such as concurrent demand, activation slope, regulatory margin, and recovery are framework-level constructs informed by established physiological research. They provide a vocabulary for organizing observable patterns across regulation, activation, and recovery without implying that each construct corresponds to a single established physiological process. This shift also changes how certain forms of openness and engagement are interpreted. Qualities often understood as stable traits—including flexibility, exploratory engagement, curiosity, reciprocity, or tolerance for novelty—may fluctuate with the regulatory conditions currently available within the system.

What follows names the core dimensions through which this organization can be read.

For the Reader

What this book is. This book presents a conceptual clinical framework grounded in converging evidence from neuroscience, developmental psychology, and embodied cognition. It proposes a reading framework through which bodily experience, regulation, and relational experience can be understood in clinical and educational contexts. Rather than offering an exhaustive empirical review, it places scientific literature in dialogue with lived experience to support a coherent reading of how experience is organized through the body.

This framework describes the conditions under which responses emerge, narrow, or become physiologically unavailable. It does not remove responsibility, but clarifies the conditions under which responsibility can be enacted.

It does not prescribe protocols or intervention strategies. Reading the nervous system does not replace intervention approaches. It helps clarify the conditions under which any approach becomes possible.

This framework does not provide a fixed method for interpreting behavior, because similar outward behaviors may emerge from very different physiological conditions. Its purpose is not to eliminate uncertainty, but to organize observation: to clarify what variables may be shaping physiological organization and how these conditions shift across time and context.

Terms such as concurrent demand, activation slope, regulatory margin, recovery, prediction, and protective organization draw on established physiological and relational research while functioning here as part of the framework's observational vocabulary. They should not be read as one-to-one equivalents of established physiological mechanisms.

This project is structured in two volumes. The first, presented here, develops the reading framework. The second, Reading the Nervous System: The Clinical Workbook, translates that framework into applied practice. The two volumes are designed to be used in sequence, but this volume stands on its own.

This volume focuses primarily on the environmental dimension of load — sensory, cognitive, social, and relational demands — as the primary terrain of observation in clinical and educational contexts. Internal physiological conditions, including unresolved pain, illness, incomplete sleep, and undetected hunger, are named where relevant; their fuller assessment is developed in the companion volume.

The text is written for professionals in health, education, and related fields seeking a reflective, situated approach to reading embodied experience. External observation remains partial; whenever accessible, first-person experience remains essential to understanding how physiological organization is actually lived.

How to read this book. The framework is built from the ground up. The first four chapters establish the parameters used throughout the rest of the book without reintroduction. Readers who begin at Part I and proceed sequentially will encounter each term as it is defined. Those who begin elsewhere—Chapter 8, 10, 11, or 12—will encounter these terms as assumed vocabulary.

The four parameters are introduced in:

- Concurrent demand — Chapter 2 (§2.3)

- Activation slope — Chapter 3 (§3.2)
- Regulatory margin — Chapter 3 (§3.1)
- Recovery — Chapter 4 (§4.1–4.9)

The Glossary provides concise definitions for quick reference. These parameters interact continuously rather than sequentially.

Chapter 5 reads them together for the first time.

Each Part builds on the previous one: Part II assumes Part I; Part III assumes Parts I and II. Individual chapters remain intelligible in isolation, but the full argument depends on the earlier foundation.

The Glossary appears at the back of the volume, before Appendix A.

Positionality and Collaboration

This book is written by a bilingual speech-language pathologist who is not herself autistic. Its framework draws substantially on published first-person autistic accounts, treated here as evidence of lived organization rather than illustration.

This creates a structural limitation that shapes what can and cannot be seen: external observation remains incomplete, and what is known from within cannot be fully recovered from outside. This principle applies to this text as it does to any clinical encounter.

The framework has not been co-developed with autistic collaborators. This shapes what it can and cannot see. Co-development with autistic researchers and clinicians would likely alter the questions asked, the constructs selected, and the weighting of evidence. The Research Agenda names this as a methodological direction.

What this text offers is disciplined attention to published autistic accounts, read through the framework presented here, and a consistent distinction between inference and established evidence. The gap between reading from autistic accounts and building with autistic collaborators remains, and is acknowledged as an ongoing responsibility.

Organizing the Regulatory Field

The framework developed in this book describes nervous system organization through a small set of interdependent parameters. These are descriptive dimensions that organize how underlying processes co-vary in lived experience.

Before entering the parameters, consider a simple moment: a person steps into a crowded hallway. Heart rate may already be elevated from the preceding class, the fluorescent hum has been present since morning, and a social exchange is expected within seconds. These are not separate events. They are one field, already in motion.

Four parameters function as the primary organizing axes of that field. Each can be considered independently, but none operates alone. They do not unfold sequentially, but as interacting aspects of a single physiological organization. What follows names them separately for descriptive clarity, not because they are separate processes.

Concurrent demand: the total processing load already present when a new demand arrives. Contributing processes: simultaneous engagement of sensory, cognitive, interoceptive, and relational processing.

Activation slope: the rate at which activation rises once recruitment begins. Physiologically, changes in activation may be reflected across autonomic activity, metabolic expenditure, and neural processing.

Regulatory margin: the available range within which activation can rise without exceeding what the system can sustain under its current conditions.

Recovery: the rate and extent to which activation moves toward a more sustainable baseline following demand. This process may involve parasympathetic re-engagement alongside changes in autonomic tone, hormonal activity, muscular tone, respiration, and cortical arousal. Considered together, three additional properties can be examined:

Threshold proximity: how close the system is operating to that threshold.

Internal time: the pacing at which activation unfolds and resolves.

Residue: activation that remains active after incomplete recovery. When unresolved, it produces baseline elevation — an altered starting condition for what follows.

The parameters described above are framework-level constructs grounded in observable patterns of regulation. Existing physiological and behavioral measures may capture aspects of these patterns, but they should not be treated as direct measures of the constructs themselves.

Table 1. Candidate Observations and Measures Relevant to the Core Constructs

Framework parameter	Physiological processes potentially relevant	Candidate observations or measures
Concurrent demand	*Simultaneous engagement across sensory, cognitive, emotional, relational, interoceptive, and other internal physiological processes drawing on shared regulatory resources. The observations below capture only a subset of these channels.*	EDA under background sensory or cognitive load, as a partial and non-specific indicator; dual-task performance cost, as a behavioral rather than physiological observation.
Activation slope	*Changes in autonomic activity, metabolic expenditure, and neural processing associated with the rise in activation.*	Heart rate rise time and rate of increase in skin conductance, as candidate measures of autonomic components of the activation trajectory.

Framework parameter	Physiological processes potentially relevant	Candidate observations or measures
Regulatory margin	*Regulatory margin is a framework-level construct describing the available range within which activation can rise without exceeding what the system can sustain under current conditions. No single physiological measure is proposed here as a direct measure of regulatory margin.*	Measures relevant to regulatory flexibility, such as HRV, may contribute contextual physiological information but should not be interpreted as direct measures of regulatory margin.
Recovery	*Physiological changes associated with recovery may include parasympathetic reengagement alongside changes in hormonal activity, muscular tone, respiration, and cortical arousal.*	Heart rate recovery and HRV recovery trajectories can index autonomic components of recovery.

They offer four observational vantage points on a single regulatory field—each a distinct clinical entry into the same process.

PART I — The Nervous System and the Construction of Bodily Experience

Part I begins at the level of shared physiological organization. Before examining variation in how nervous systems organize experience, it establishes the regulatory processes they hold in common.

This section describes how nervous systems manage activation, carry demand, and restore availability in continuous interaction with the environment. These processes are presented without reference to any specific population, as properties of nervous system organization itself.

What is described here is not a special case. It is the condition from which all variation emerges.

Chapter 1 — How the Body Builds an Inner World

A sound arrives. The body registers it. Breath shifts slightly. Attention moves. The system is adjusting as experience forms.

Before behavior, before named emotion, before language organizes meaning, a nervous system is at work. It receives signals, weighs them, integrates them across domains, and updates its expectations about relevance as conditions change.

What follows is a way of reading regulatory organization across time.

At the level of physiology, this is activation, recovery, and what recovery leaves behind. At the level of experience, it is what it feels like to be present and available — or to arrive at a moment already carrying more than it can hold.

What that adjustment leaves behind becomes the ground of whatever follows. Experience does not begin from zero. It begins from what previous organization has already made possible — or foreclosed.

1.1 — What Is Being Organized?

To say that the body organizes experience is a physiological description, not a metaphor. At every moment, the nervous system is integrating multiple streams of signal: input arising from within the body, input from the external environment, patterns carried forward from prior activity, and the actions available in the present. What is being organized is this continuous flow. Signals arising from within the body—interoceptive signals—provide ongoing information about internal state: changes in heart rate, respiration, muscle tone, temperature, visceral activity. These are not background conditions. They are part of the material from which experience is formed. At the same time, exteroceptive input—sound, light, movement, contact—enters a system that is already active, already carrying prior configurations of activation and resolution.

Organization occurs as these streams are brought into relation. The system does not process each signal independently. It organizes across them — shaping what the system can perceive, what it can sustain, and what actions the current state makes available. Prior activity does not disappear once an event ends. What has not fully resolved remains active, carried forward as part of the baseline from which the next moment is met.

Before a word is spoken, the body has already organized posture, breath, orientation, and readiness for action. These are not separate processes. They are dimensions of a single ongoing organization.

To read the nervous system is to follow how these signals are being organized across time — how states take shape, how they are sustained, and how they resolve or remain active into what follows.

1.2 — The Nervous System as a Predictive and Selective System

Multiple streams of signal arrive simultaneously, and what the system carries into each moment reflects both what is present and what prior processing has left unresolved.

This book adopts a predictive-processing framing — one of several current models of nervous system function — in which the nervous system operates as a predictive system, continuously generating expectations about the body and the environment and comparing those expectations with incoming signals (Seth et al., 2012; Seth, 2013).

This adjustment is not localized. Autonomic activity, muscular tone, and internal signaling shift in parallel as the system responds to incoming conditions.

These responses are shaped by bodily memory—patterns of physiological response formed through repeated experience that influence how present signals are interpreted. Each moment is encountered through this accumulated history of prior regulation.

Prediction errors arise from two physiologically distinct sources. Exteroceptive prediction errors occur when sensory input from the environment diverges from expectation—an unexpected sound, a sudden movement, or a change in context. Interoceptive prediction errors occur when signals from within the body diverge from what the system anticipated—a shift in heart rate, an unexpected visceral signal, or an altered pattern of breathing.

Both recruit adjustment and shape how much the system is already carrying and how quickly activation begins to climb. The distinction matters because systems vary in how precisely they track each channel — and how steeply activation rises when each fails to predict accurately. This variation becomes central in Part II.

At every moment, more signals are present than the system can process. The nervous system therefore continuously manages relevance. Not all signals are treated equally. A system that attempted to process every incoming input with equal weight would be overwhelmed within seconds.

This selective management — how the system determines what continues to require active processing and what no longer does — is developed in Chapter 2.

The nervous system is therefore both predictive and selective—simultaneously generating expectations and regulating which signals require active response. Together, these processes determine what remains active and what recedes.

This selective organization is where nervous systems begin to differ in ways that matter. Thresholds for treating signals as relevant, the speed with which repeated input is down-weighted, and the precision with which prediction tracks particular domains all vary. They reflect differences in architecture, experience, and prior learning. Contemporary predictive accounts likewise propose that perception emerges through ongoing prediction and updating rather than passive reception (Clark, 2013).

The nervous system does not respond to events as fixed units. It responds to the signals those events generate, and to the signals the body is already carrying. What appears to be a reaction to something that has happened is often the continuation, amplification, or resolution of signal configurations already in motion. From this level of reading, the same event does not exist as a stable object to which the system responds, but as a variable set of conditions whose meaning is shaped by the state in which it is encountered.

The event is one. The signals are many. The response follows the signals. Figure 1 represents this prediction–comparison loop.

Figure 1 — Prediction–Comparison Loop in Nervous System Function

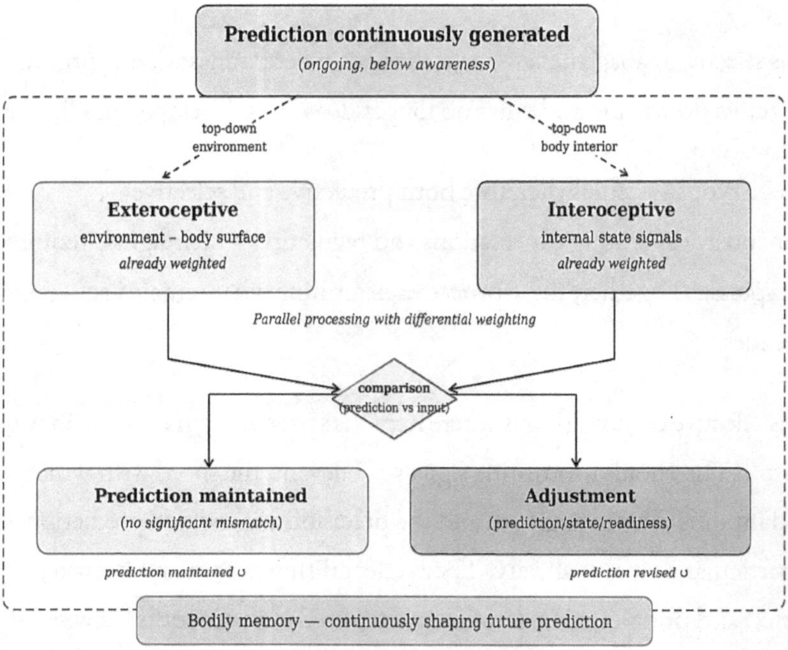

This figure is not a model to be applied. It is a simplified reading of a process that is already occurring continuously.

Reading the Nervous System • María E. Reyes, M.S., CCC-SLP

This figure is read as a continuous loop, showing how predictions are generated, compared with incoming signals, and either maintained or revised as conditions change. Each cycle begins from the state the system is already carrying.

1.3 — Baseline Organization

The baseline state is the condition the system is in—the ground from which activation, response, and recovery proceed (Sterling, 2012; Ramsay & Woods, 2014). It is not a strategy, but the condition that makes behavior possible.

In many contexts, regulation is treated as something a person must achieve before they can learn, engage, or communicate—as though it were a skill acquired through effort. Here, regulation refers to something more fundamental: the ongoing coordination of multiple physiological systems across time that allows the organism to remain coherently present.

Even in stillness, the system is actively regulating internal signals—adjusting autonomic tone, weighting sensory input, and maintaining the conditions that allow orientation to continue. There is no neutral state from which regulation begins; there is only the state the system is in.

A task is given. The body is already within a state. If the state supports integration, the task is taken up. If not, it remains out of reach. Nothing about the task has changed. The condition of the system has.

At rest, a nervous system maintains a patterned stability that allows activation to rise and fall without loss of coherence. This stability is not static; it adjusts continuously to surrounding conditions.

This organized stability allows the system to continue responding coherently to the demands it encounters. It is not the result of those responses; it is the condition from which they become organized (Sterling, 2012; Ramsay & Woods, 2014).

What appears as capacity is the expression of a state already supporting it.

When a nervous system struggles to sustain stable organization under increased demand, the relevant process is physiological before it is behavioral. The question is therefore not whether the person is trying, but what the system can currently organize. The primary task of the nervous system is to maintain viable continuity — to keep the organism sufficiently organized to remain present (Sterling, 2012).

This regulatory foundation makes visible four parameters that guide the chapters that follow. First, what the system is carrying before new demands arrive. Second, how steeply activation rises when predictions fail. Third, the space available between current activation and the limits of organized functioning. And fourth, how completely the system returns toward baseline. These are examined throughout Part I as interdependent dimensions of the regulatory field.

What follows traces how this baseline shifts under demand.

1.4 — What the Physiology Reveals

What becomes visible are the conditions that produce behavior—prior to behavior itself.

The system is always carrying something before any new demand arrives. Processing does not begin at zero, but from the state produced by prior activity, the current environment, the history of regulation across time, and what the body itself is carrying internally. That starting condition shapes what remains available for whatever follows.

As input diverges from expectation and activation begins to rise, the physiological cost cannot be inferred from the size of the input alone. The same input may be met differently depending on the state from which it is received.

What follows an event is therefore as important as what precedes it. The system does not necessarily meet each new moment from the same starting point.

These are not four separate observations. They are the same regulatory field read from four different vantage points. Each chapter that follows takes up one of these vantage points in turn. Figure 2 traces this across a single regulatory cycle.

Figure 2 — Continuous Dynamics of Activation Across Time

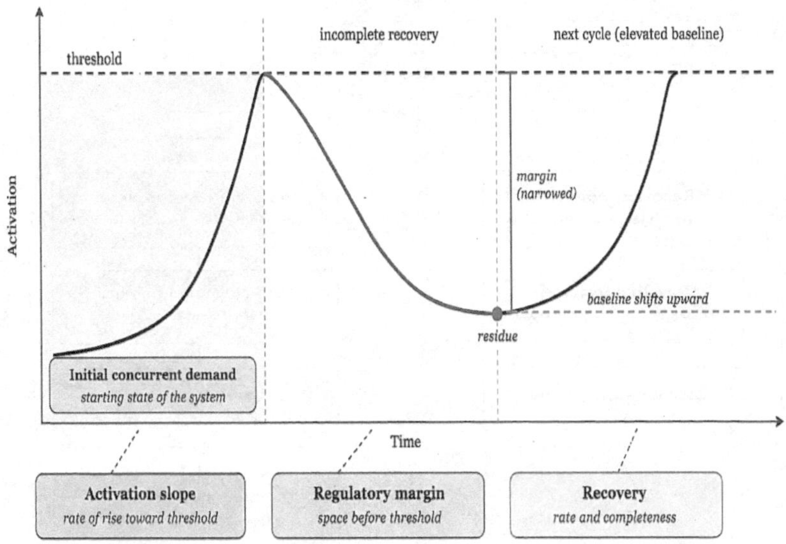

This figure represents the nervous system as a continuous regulatory process unfolding across time. Each cycle begins from an existing baseline – the concurrent demand – reflecting prior activity, environmental conditions, and incomplete resolution of earlier activation.

This figure is not a model to be applied. It is a simplified reading of a process that is already occurring continuously.

Reading the Nervous System • María E. Reyes, M.S., CCC-SLP

Continuous movement of activation across a single regulatory cycle. This figure is read from left to right, showing the rise of activation in response to demand and its return toward baseline under varying recovery conditions. Figure 3 maps this across repeated cycles.

Figure 3 — Regulatory Cycle Across Time

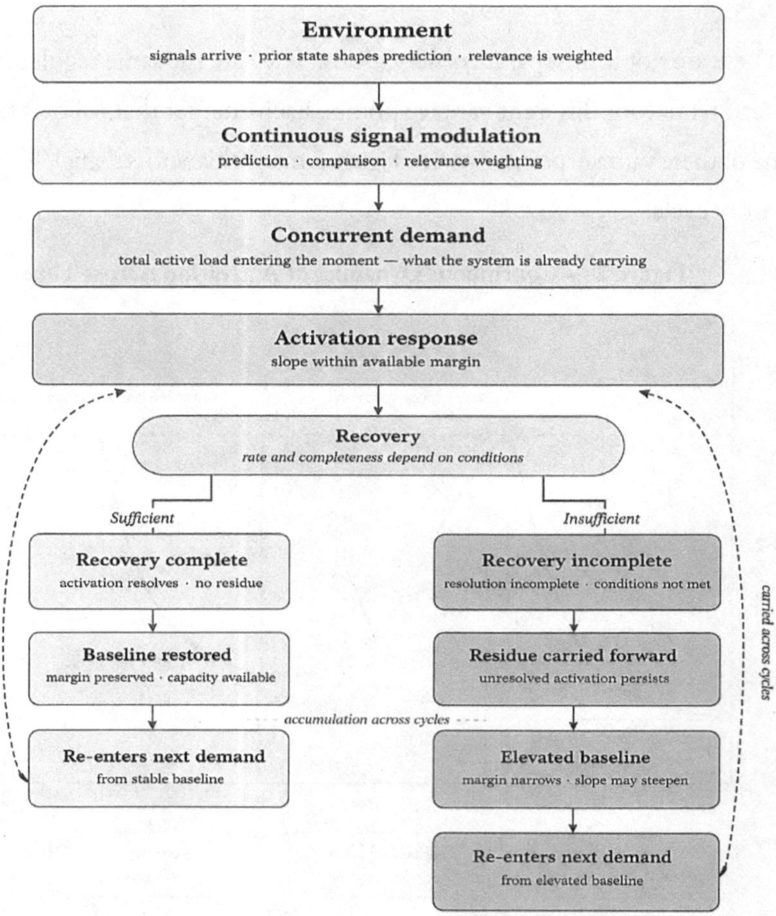

The nervous system is not responding to events, but continuously reorganizing across cycles depending on what was carried forward.

This figure is not a model to be applied. It is a simplified reading of a process that is already occurring continuously.

Reading the Nervous System • María E. Reyes, M.S., CCC-SLP

Recursive regulatory cycles in which each cycle begins from the state the previous one leaves behind.

This figure is read from top to bottom as a continuous regulatory cycle unfolding across time. The two pathways represent divergence across repeated cycles depending on whether sufficient recovery becomes physiologically possible.

The lighter pathway illustrates sufficient recovery: activation resolves, baseline returns, and the space available for further activation remains open across subsequent cycles. The denser pathway illustrates accumulation: what remains unresolved carries forward, baseline progressively elevates, and the system re-enters each new cycle with less space available before reaching its limits.

This is what the chapters that follow develop: the conditions the system is already managing, how much margin remains available, how activation shifts when a new demand arrives, and what happens to what does not finish resolving.

Before anything happens, something is already being carried.

Chapter 2 — What the System Is Already Carrying

It is early morning.

Two people arrive at the same building within minutes of each other. The same entrance. The same corridor. The same fluorescent lighting cycling at its characteristic frequency. The same ambient sound from the ventilation system. The same cluster of colleagues near the coffee station speaking at normal conversational volume. The same smell. The same visual complexity of a busy open-plan floor settling into its morning rhythm.

From the outside, both have arrived. The environment is identical.

The morning has not yet made any demand.

But the nervous systems that have just walked through that door are not receiving the same morning.

For the first person, the fluorescent hum registers briefly on entry and settles into background within the first minute — noted, filed as non-informative, no longer requiring attention. The visual field of the floor organizes into a stable pattern that does not require further active management. By the time this person reaches their desk, the environment has largely resolved. The task ahead is what remains active.

For the second person, the fluorescent hum takes a little longer to settle — it is still faintly present at the desk, not distressing, not consciously attended to, but occupying a small amount of available processing. The visual field requires a second pass before it stabilizes. By the time this person reaches their desk, the environment has mostly resolved, but not entirely. A small fraction of what just arrived is still being managed.

Neither person is struggling. Neither is doing anything wrong.

What differs is not the environment and not the effort. It is the pace at which repeated input settles into background — and how much of the environment each system is still actively managing before the first task begins.

The room is the same. The load is not.

Contemporary neuroscience describes perception as an active process in which the nervous system continuously predicts incoming signals and adjusts when those predictions diverge from sensory input. Differences in how signals are weighted, suppressed, or sustained shape how much processing demand a given environment continues to generate—and for how long after initial entry. This chapter begins here—not when something goes wrong, but in the ordinary divergence of nervous systems encountering an environment built around the assumptions of one of them.

2.1 — The System Is Never at Zero

That divergence begins before any demand is placed—in the baseline state from which the day's first moment is received.

Each moment begins from wherever the preceding period has left the system — shaped by what was processed, what was suppressed, what was resolved, and what is still being managed. This includes not only what the environment has demanded, but what the body itself is carrying: unresolved pain, illness, incomplete sleep, undetected hunger. These are not background conditions. They are active inputs into the baseline from which every subsequent demand is received.

Baseline, as used here, does not refer to a theoretical resting state measured under controlled conditions. It refers to the system's actual starting condition: the activation it carries forward, the processing resources already committed, the predictions already in motion, and the signals still being managed.

Every new demand arrives into this inherited state rather than an empty system.

The load a system carries is determined not only by what is present, but by how long prior activation has remained unresolved. Demand is not reset between events; it is carried forward until resolved.

2.2 — Habituation and the Management of Signal

One way the nervous system manages baseline load is at the level of signal—through processes of active suppression.

The nervous system receives far more input than it can process with full attentional resources. This is the fundamental condition of being a sensory organism in a complex environment. The regulatory task is to determine what must be processed and what can be released.

Habituation refers to a reduction in response following repeated exposure to a stimulus. It is not simply passive fading, and multiple neural and learning processes may contribute to it. When repeated input becomes predictable and carries little new relevance, responding may decrease, allowing less processing to be devoted to that input while attention remains available for what is unresolved.

Classic habituation research demonstrated that repeated, non-informative stimulation can produce diminished responding over time (Thompson & Spencer, 1966). Within the present framework, more complete habituation is treated as one condition that may reduce ongoing processing demand, whereas persistent responsivity may contribute to continued demand even when the stimulus does not reach conscious awareness.

The rate and completeness of habituation vary across individuals and contexts. RNS treats that variability as one possible contributor to differences in how long environmental signals continue to carry regulatory cost.

Habituation research shows that responding to repeated stimulation can diminish over time and that this process varies across organisms and conditions (Thompson & Spencer, 1966). In RNS, these findings inform—but do not by themselves establish—the framework's account of how repeated or predictable sensory input may recede in processing weight.

This difference changes how much processing work is already underway when the next demand arrives. A nervous system that suppresses signals quickly carries less environmental processing into the next task. One that suppresses signals more slowly continues to manage them as the task unfolds.

This variability will become especially relevant in Part II, where we examine how these same mechanisms contribute to understanding the physiological organization of autistic experience.

Figure 4 shows how incoming signals are distributed based on this process.

Figure 4 — Physiological Fates of Signals After Relevance Weighting

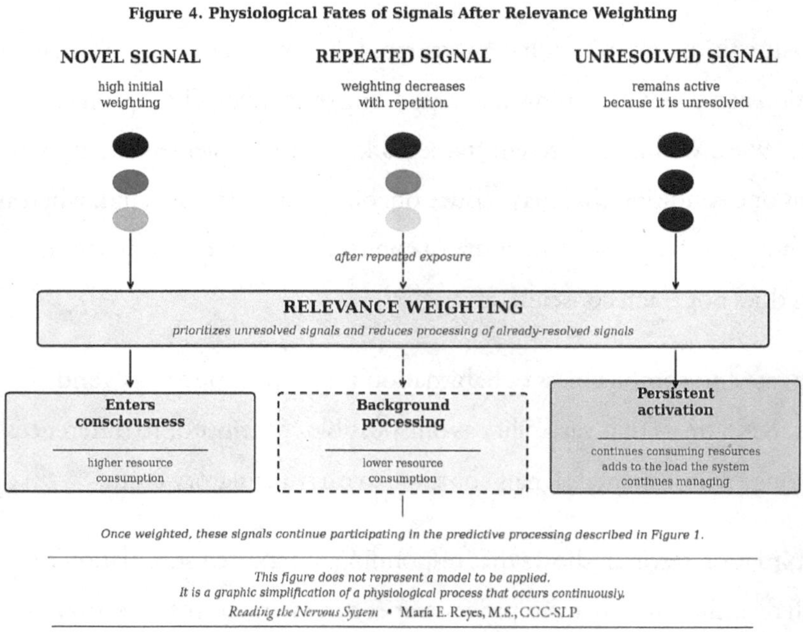

2.3 — Concurrent Demand

Concurrent demand encompasses more than the immediately presenting task.

A nervous system engaged in a task is not managing only that task. It is also processing the surrounding sensory environment, monitoring internal states, sustaining postural organization, tracking social cues, maintaining predictions about what may occur next, and suppressing signals that remain unresolved.

All of this occurs at the same time. These are not conditions that modify response. They are sources of demand in themselves, operating through the same mechanisms as external load. A body managing an internal physiological condition enters every environmental demand already loaded.

Visible effort is not a reliable indicator of load. A person may be managing high concurrent demand while appearing calm and competent. Successful management often produces little outward sign. What appears easy may be metabolically costly. The absence of visible difficulty does not indicate low load, but that the system is managing that load within its available capacity.

Two systems can retain the same capacity and still produce different outcomes under different conditions. They can also produce the same observable outcome while carrying very different physiological loads. What changes is not necessarily the behavior. What changes is the physiological organization required to produce it. Part of that load resolves before the next demand arrives; part remains active and becomes part of the conditions under which the system meets what comes next.

Concurrent Demand is not proposed here as a newly discovered physiological mechanism. Research across attention, workload, stress, and person–environment models has long demonstrated that functioning depends on the demands placed on limited and interacting resources. RNS uses the term Concurrent Demand as a clinical organizing construct: a way of attending to the sensory, interoceptive, cognitive, relational, postural, and contextual demands that may be operating simultaneously in a given moment.

Figure 5 — Divergent Concurrent Demand Across Time in a Shared Environment

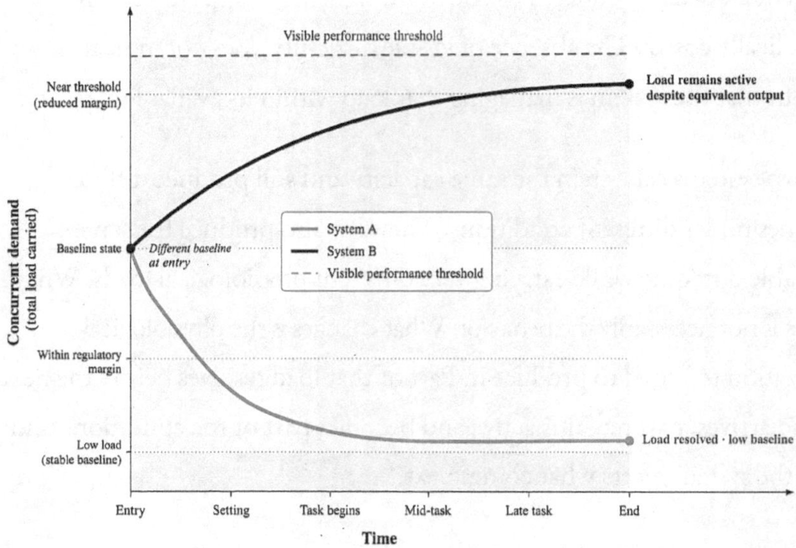

Both systems remain below the visible performance threshold; output appears equivalent while concurrent demand diverges across time.

This figure is not a model to be applied. It is a simplified reading of a process that is already occurring continuously.

Reading the Nervous System • María E. Reyes, M.S., CCC-SLP

2.4 —The Invisible Cost of Competent Management

When a nervous system manages high concurrent demand successfully, the demand itself disappears from view.

Observers see participation, responsiveness, and appropriate behavior. The work required to produce those outputs remains internal to the system and largely invisible.

This invisibility produces a consistent misinterpretation. When a child completes a demanding school morning without visible difficulty, it is assumed that the morning imposed little demand. When an adult participates in a full day of meetings without disruption, it is assumed that the day was manageable.

These interpretations confuse output with cost.

A person reading only visible behavior would note engagement and conclude the task went well. A person reading concurrent demand would ask what the system was managing before the task began, and what it will carry into the next hour.

What is not visible is still being paid.

Across longer timescales, this kind of accumulated regulatory effort may contribute to what the allostatic-load literature describes as the cumulative biological cost of maintaining stability through repeated adaptation to demand (McEwen & Stellar, 1993; Sapolsky, 2004; Sterling, 2012). Concurrent demand describes the moment-to-moment conditions under which that cost may accumulate.

A system that appears fine at the end of the morning may arrive at the afternoon already carrying what that period required. The cost does not disappear with the morning. It remains as what the morning left behind.

Disruption later in the day is therefore frequently misread.

Micro-Sequence — The Invisible Morning

Eight fourteen. The door opens. Sound, light, movement, proximity—the system begins assessing. Some signals are suppressed within seconds. Others remain active.

The desk is reached. The task begins. The environment continues arriving.

Two hours pass. The output is normal. Participation appears effortless. The baseline has shifted. No visible signal marks the shift.

A small unexpected change occurs. The response is larger than the change appears to warrant. From the outside, it appears disproportionate.

Within an RNS reading, one possible interpretation is that the visible response reflects accumulated cost from earlier in the morning that was not apparent at the behavioral surface. Figure 6 traces this accumulation beneath equivalent output.

Figure 6 — Equivalent Observable Outcome With Progressive Accumulation of Physiological Load

Figure 6. Equivalent Observable Outcome With Progressive Accumulation of Physiological Load

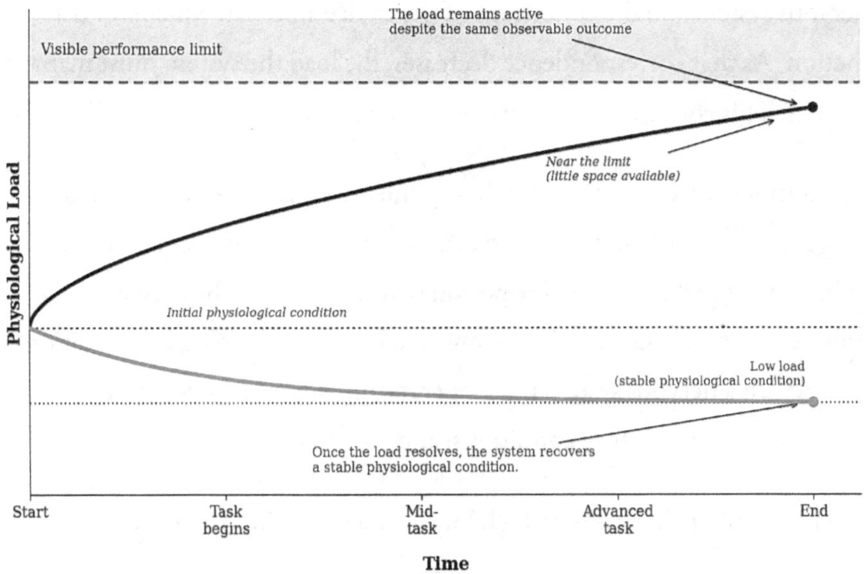

Both trajectories remain below the visible performance limit; the outcome appears equivalent while physiological load diverges over time.

This figure is not a model to be applied. It is a simplified reading of a process that is already occurring continuously.

Reading the Nervous System • María E. Reyes, M.S., CCC-SLP

2.5 — Mismatch

Concurrent demand is not a property of a person, and it is not a property of an environment. It arises from the relationship between a nervous system's processing architecture and the environment within which it is required to function.

An environment designed around the habituation thresholds and processing speeds of one kind of nervous system will impose relatively low concurrent demand on systems that match those assumptions, and higher demand on those that do not.

This is mismatch.

Mismatch names the lack of correspondence between the characteristics of a nervous system and the conditions of the environment in which it must function. As that correspondence decreases, the load the system must manage increases. This concept will be developed further in Part II.

Relational accounts of autism have similarly emphasized that difficulty can emerge through mismatch between differently situated people rather than residing solely within the autistic person (Milton, 2012). The present framework extends the question of mismatch to the physiological conditions under which a nervous system must operate, including sensory demand, processing pace, regulatory conditions, and recovery.

This framing changes how high load in a setting is understood.

When it is interpreted as a problem within the individual, attention remains on behavior. When it is understood as the product of mismatch between the system and the environment, attention shifts toward the conditions generating that load.

The question stops being "What is this person doing wrong?" and becomes: what signals remain active? what environmental demands require sustained processing? what features of the context are contributing to the load the system is already carrying?

These questions locate the problem at the level where it occurs.

What concurrent demand determines is not only the current state of the system, but the conditions under which every subsequent demand will be received.

2.6 — What Is Activation

When the system registers a demand that requires reorganization, it responds. That response is activation.

Activation is the coordinated physiological shift that occurs when the system registers a demand: heart rate rises, respiratory patterns change, muscular tone increases, cortical arousal shifts. These changes reflect the rise in the system's physiological activation in response to the load it is managing.

Activation does not, by itself, constitute a sign of dysfunction. It constitutes the physiological expression of a system responding to the load it is managing.

Chapter 3 — The Dynamics of the Response

The meeting has been running for forty minutes.

Both people are present, engaged, and managing the demands of the room—tracking the conversation, preparing responses, monitoring relational cues, and sustaining the composure the context requires. From the outside, both appear to be participating in the same way.

Then the agenda shifts without warning. A topic scheduled for next week is introduced now. The change is brief—a redirection, a new set of materials placed on the table—and the meeting continues without pause.

For the first person, the shift registers and is absorbed. The new topic is taken up. Activation rises slightly and begins settling almost immediately. Within a minute, the system reorganizes around the revised agenda and returns to a functional state.

For the second person, the prediction failure generates a steeper activation arc. Activation increases more quickly as the system attempts to reorganize around the shift. The prediction that had been organizing the next hour of the meeting has been invalidated. The materials mentally prepared for next week are no longer relevant. The social meaning of the shift—why now, what it requires, whether the preparation holds—requires assessment that cannot complete in the seconds the room allows. The conversation continues while the system is still reorganizing its response.

A breath held slightly. Attention narrowed onto the new materials while still processing the transition. The next sentence arrives in the room before the system has finished absorbing the previous one. This does not imply distress. It is the activation curve still rising while the surface remains composed.

From the outside, nothing appears different. Both people remain in the meeting.

Inside the second system, however, the unexpected shift has produced a faster rise in activation. Although both people continue participating in the meeting, the second system is still reorganizing its response as new demands keep arriving.

3.1 — Margin Is Never Fixed

Regulatory margin is not a fixed quantity. It describes how much space remains available for the system to absorb new demands without activation exceeding what it can sustain under its current conditions.

As concurrent demand rises, baseline rises and margin narrows. A system that began the morning with wide margin may arrive at a later demand with very little.

This relationship becomes visible across baseline and threshold.

Figure 7 — The Threshold Has Not Moved. The Baseline Has Risen Toward It.

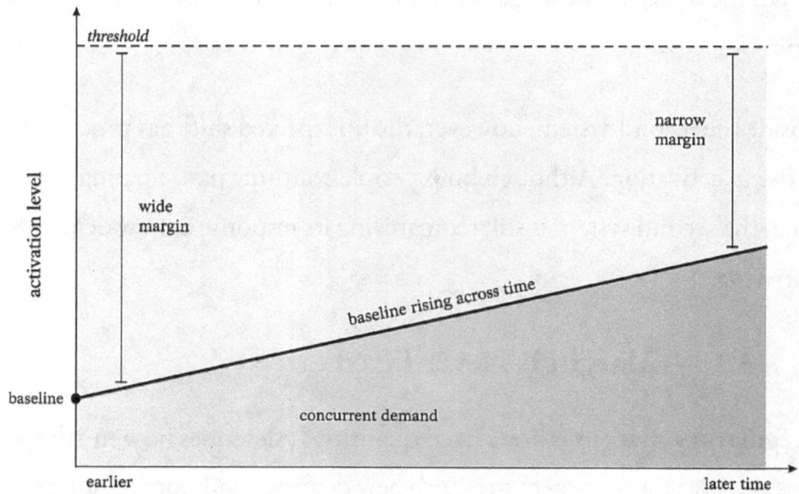

Regulatory margin is the distance between current activation and threshold.
As baseline rises across time, that distance narrows — even when threshold remains unchanged.

This figure is not a model to be applied. It is a simplified reading of a process that is already occurring continuously.

Reading the Nervous System • María E. Reyes, M.S., CCC-SLP

The distinction matters operationally. Margin describes the system's remaining range to take on new demand without rising activation exceeding what it can sustain. It does not describe how much activation exists in absolute terms, but how much space remains available from the system's current physiological organization.

This is why the same perturbation produces different outcomes at different moments. When margin is wide, the system can absorb a larger rise in activation without leaving the range it can sustain. When margin has narrowed, the same rise finds less available space.

At the level of observation, narrowing margin becomes visible through reduced flexibility under increasing demand, increased effort to sustain the same output, and earlier transition into protective organization across comparable contexts.

Regulatory margin should not be inferred from behavioral success or difficulty alone. A person completing a demand does not establish that margin was wide, just as difficulty completing it does not establish that margin was narrow. Within this framework, margin is inferred cautiously from converging patterns across comparable conditions: changes in demand, activation trajectory, effort, recovery, access, self-report when available, and what changes when conditions are adjusted. No single behavioral outcome identifies regulatory margin directly.

A system managing high unsuppressed signal across a demanding morning arrives at a new perturbation with less margin than it had earlier in the day.

Margin meets a critical point at threshold, where rising activation exceeds what the system can sustain under its current conditions. Most rises in activation remain within available margin, absorbed before they reach that point.

A system with low concurrent demand and wide margin can sustain a larger rise in activation before threshold is reached. A system already managing high load reaches it sooner. What changes is not the signal, but the margin left to manage it—and this can appear, from the outside, as increasing sensitivity across the day (Thayer & Lane, 2000): the inputs have not necessarily changed; the system's margin has.

Environments carry implicit assumptions not only about how much signal is present, but about how much margin they assume the system has available to absorb it. For systems already operating with reduced margin, that assumption does not hold. This is mismatch expressed at the level of margin.

3.2 — The Slope

Activation rises along a curve within the available regulatory margin.

The steepness of that curve—activation slope—describes how quickly activation rises. The threshold marks the point at which that rising activation exceeds the available regulatory margin and begins to recruit protective reorganization.

Slope describes the speed of the rise. Margin describes how much of that rise the system can sustain.

What changes across systems is not whether activation occurs, but how quickly it unfolds.

The same external demand may produce different slopes depending on the concurrent demand already present. The same slope may consume different amounts of regulatory margin depending on what the system is already carrying. The perturbation still matters—but the observable response reflects how it interacts with the system's activation dynamics at that moment, not the input size alone.

A shallow slope produces a more gradual rise. A steep slope carries activation to higher levels in less time. The difference does not necessarily lie in the size of the perturbation, but in the speed at which activation changes once the response has begun.

This relationship becomes visible when slope and margin are represented together. Figure 8 shows two trajectories from the same perturbation.

Figure 8 — Activation Slope: Divergent Trajectories Following the Same Perturbation

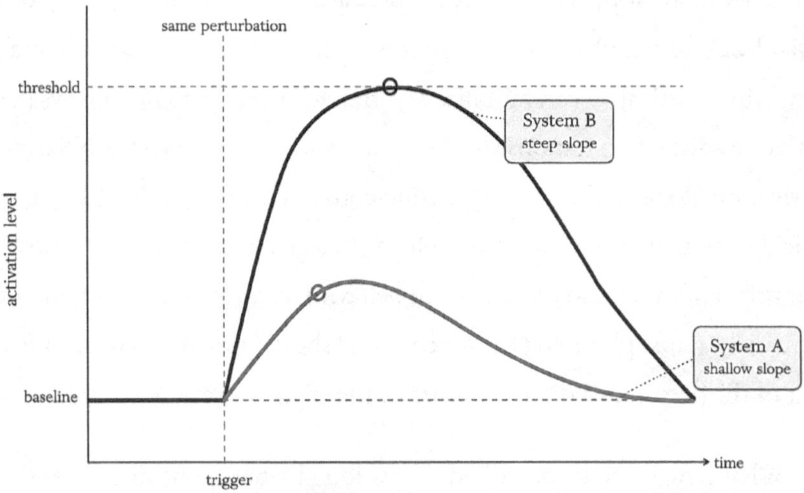

The same perturbation produces different activation trajectorios depending on the system's slope.
A steeper slope generates a faster rise and higher peak, independent of the trigger itself.

This figure is not a model to be applied. It is a simplified reading of a process that is already occurring continuously.

Reading the Nervous System • María E. Reyes, M.S., CCC-SLP

In physiological terms, activation may involve measurable changes in autonomic activity, metabolic expenditure, and neural processing as the system mobilizes to manage a perturbation.

These physiological changes and visible behavior do not necessarily unfold on the same timeline.

A person may remain externally composed while internal activation rises steeply. Within this framework, physiological activation and visible behavior are understood to operate at different levels — a distinction long recognized in research on the dissociability of autonomic, endocrine, and behavioral indices of arousal (Lacey, 1967; Porges, 2011); a steep activation slope may therefore remain difficult to infer from outward behavior alone.

Activation Slope is not proposed here as a single measured physiological variable. Different physiological systems — cardiovascular, electrodermal, endocrine, cortical — can mobilize at different speeds, and not always in step with one another, in response to the same event (Lacey, 1967). RNS uses Activation Slope as a clinical organizing construct: a way of attending to how quickly a system appears to be mobilizing in a given moment — understanding that this cannot be read directly from behavior, cannot be assumed to unfold identically across physiological systems, and should not be treated as a fixed trait of the person or a direct measure of the triggering event.

What remains to be described is how long that arc continues after the signal that generated it has passed.

3.3 — Momentum

Activation does not necessarily end when the signal that triggered it disappears. Once begun, it can continue unfolding across time even after the initial stimulus has passed. This continuation is momentum.

Slope and momentum describe different properties of the activation curve. Slope describes how quickly activation rises. Momentum describes how long it continues unfolding once the response has begun. Activation can rise quickly and begin settling soon after, or rise more gradually and extend for longer. For this reason, the speed of the rise does not by itself determine how much momentum the response develops.

This difference becomes critical at the level of observation. By the time behavior becomes visible, the internal activation arc may still be unfolding. What is observed is not the rise itself, but the system after momentum has carried activation forward.

Figure 9 — The Activation Arc Across Time: Slope, Margin, and the Temporal Displacement of Behavior

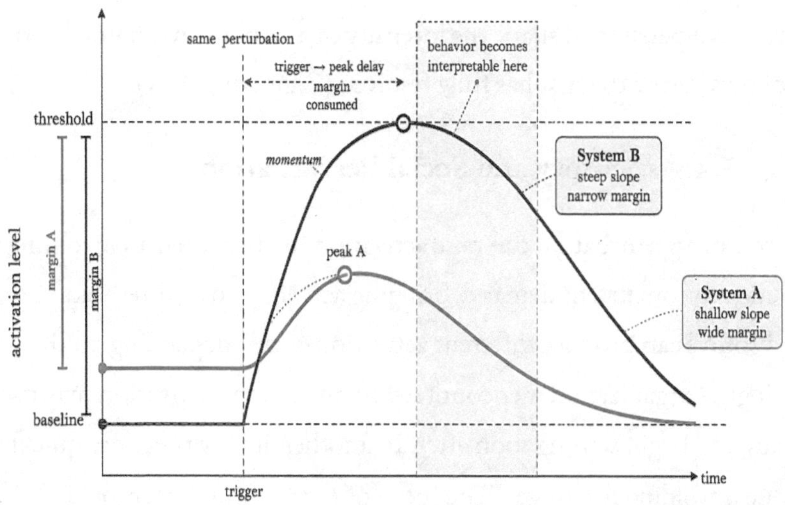

Same trigger. Different curves. Behavior becomes interpretable after the arc has already peaked.

This figure is not a model to be applied. It is a simplified reading of a process that is already occurring continuously.

Reading the Nervous System • María E. Reyes, M.S., CCC-SLP

This figure shows how identical perturbations produce different delays between trigger and peak activation, and how momentum extends activation beyond the point at which behavior becomes visible.

Momentum alters how events are interpreted. A response that appears delayed or disproportionate may reflect not the immediate trigger, but the continued unfolding of an activation arc already in motion.

The system is not responding only to the present. It is completing what has already been set in motion.

This distinction has a parallel in cardiovascular stress research, where reactivity and recovery have long been treated as separable, only loosely correlated properties of a stress response (Linden, Earle, Gerin, & Christenfeld, 1997). A related literature on residual arousal describes how activation from one event can persist and shape the intensity of a response to an unrelated event that follows before the first has fully resolved (Zillmann, 1971).

3.4 — Sensory and Social Perturbation

A sensory perturbation can raise activation further when an incoming signal adds to concurrent demand, bringing activation past threshold. A sudden sound can produce different activation curves depending on the physiological organization it encounters. In one system, activation may rise gradually and begin settling soon after. In another, it may rise more quickly or continue unfolding for longer. The speed of the rise and the temporal extension of the response describe different dimensions of those curves.

Social perturbation follows the same dynamic. The nervous system continuously predicts relational cues—tone, timing, facial expression, and conversational rhythm. When these predictions fail, activation may be recruited. Social environments generate continuous novelty and ambiguity, requiring constant prediction updating and ongoing evaluation of ambiguous information. For systems with lower thresholds for social uncertainty, or steeper activation slopes in response to ambiguity, ordinary social contexts can generate higher levels of processing demand and physiological cost.

This is mismatch again, now at the social level: environments also carry assumptions about the cost of navigating relational unpredictability, not only about the amount of sensory signal present.

What is being processed does not imply danger. It is uncertainty.

3.5 — What Determines the Slope

Concurrent demand matters because it shapes the state from which the response begins. Recent activation history and the demands the system is already managing form part of the initial condition from which the new perturbation is encountered.

The nature of the perturbation also matters. Novelty, unpredictability, and the discrepancy between what was anticipated and what was encountered can shape the reorganization required when a new demand arrives. No single one of these variables determines slope on its own.

For this reason, a steep slope should not be read as a stable trait of the person, nor as a direct measure of the size of the event.

Figure 10 — The Same Slope May Consume Different Amounts of Regulatory Margin

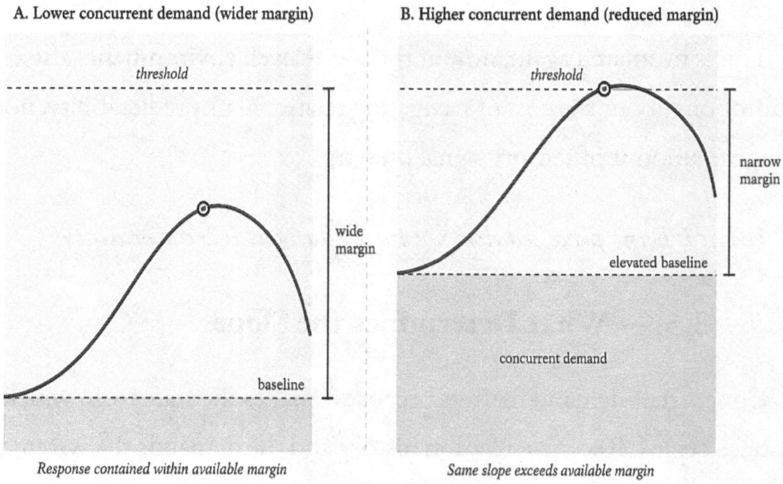

The same slope produces different outcomes depending on available margin.
When margin is wide, the response remains contained. When reduced by concurrent demand, the same slope exceeds capacity.
Outcome is shaped not only by slope, but by where it begins.

This figure is not a model to be applied. It is a simplified reading of a process that is already occurring continuously.

Reading the Nervous System • María E. Reyes, M.S., CCC-SLP

3.6 — The Behavioral Surface of a Steep Slope

A steep activation slope does not have a single behavioral expression. Activation can rise quickly while a person continues talking, working, responding, or remaining in relationship. At other times, that rise may be accompanied by shorter responses, reduced elaboration, or narrowed attention.

Visible behavior does not, on its own, allow the shape of the preceding activation curve to be reconstructed.

Different physiological curves can produce similar behavioral surfaces. Likewise, different behavioral responses can emerge from systems managing comparable rises in activation under different conditions.

For this reason, behavior is not a direct measure of slope.

What is observed is the surface of a physiological dynamic that must be read across time and in relation to the conditions in which it emerges.

What remains to be described is not how activation begins, but how it ends.

Chapter 4 — What the System Does With What It Carries

The meeting ended twenty minutes ago.

The person is back at their desk. The acute demand has passed. The conversation that generated the activation arc is over. The room is quiet. Nothing is currently being asked of the system.

And yet the system has not returned to baseline.

Breath remains slightly shallower than it was earlier in the morning. Muscle tone along the jaw and upper back has not fully released. The attentional field, narrowed during the meeting, has not yet redistributed. A sentence that would have formed in three seconds before the meeting now takes five. The work on the desk is familiar—known material, no new demand—yet it requires more effort than it did earlier.

The meeting is over. The activation is not.

The curve has peaked and descent has begun, but that descent is neither instantaneous nor complete. The system is somewhere along the descending slope—below the peak, above baseline, moving at a pace the system has not yet resolved.

How far above baseline the system remains, how long it takes to return, and whether recovery will occur before the next demand arrives—these are the questions this chapter addresses.

4.1 — Recovery Is Not Passive

Recovery is active, not passive. The conditions under which it proceeds—and the pace at which it unfolds—are governed by the system's state, not by the removal of demand.

Recovery is not goal-directed. It is the unfolding of physiological dynamics under conditions that permit return toward baseline.

The most consequential misunderstanding of recovery is also the simplest: that it occurs automatically once demand stops. In this view, recovery is passive. Remove the demand, and the system returns to neutral.

This is not how recovery operates. The absence of demand does not guarantee recovery.

Recovery—the extent to which the system returns toward baseline following activation—is an active physiological process that may involve parasympathetic re-engagement alongside broader changes in autonomic tone, hormonal activity, muscular tone, respiration, and cortical arousal. It requires specific conditions to proceed and unfolds at its own pace, governed by the state the system is in, not the state the external environment has moved to.

Recovery follows the system, not the schedule.

When demand stops and conditions permit, the system begins returning toward baseline. Autonomic tone shifts. Sympathetic activation decreases while parasympathetic activity increases. Hormonal elevations begin to clear. Muscular tone releases. Breathing deepens and slows. Cortical arousal redistributes from a narrowed configuration toward a broader, more flexible state.

Recovery unfolds as coordinated physiological change, not as the absence of stimulation.

These dynamics are consistent with research on autonomic regulation and heart–brain interaction, which describes recovery as an active regulatory process with measurable timescales rather than a simple absence of stimulation (Thayer & Lane, 2000; Shaffer & Ginsberg, 2017). None of this occurs instantly, nor is it guaranteed by the removal of demand alone.

4.2 — Rate and Completeness

Recovery varies along two independent dimensions: how quickly activation descends, and how completely it returns to prior baseline.

The first is **rate**—how quickly the activation curve descends after its peak, reflecting the speed at which autonomic arousal decreases and physiological markers move toward resting levels.

The second is **completeness**—whether the system returns to its prior baseline or settles at a higher level, carrying forward an increment of the activation cost.

Rate determines how quickly the system descends. Completeness determines where it lands. This relationship becomes visible when both parameters are shown together. Figure 11 separates these as independent parameters.

Figure 11 — Rate and Completeness: Two Independent Parameters of Recovery

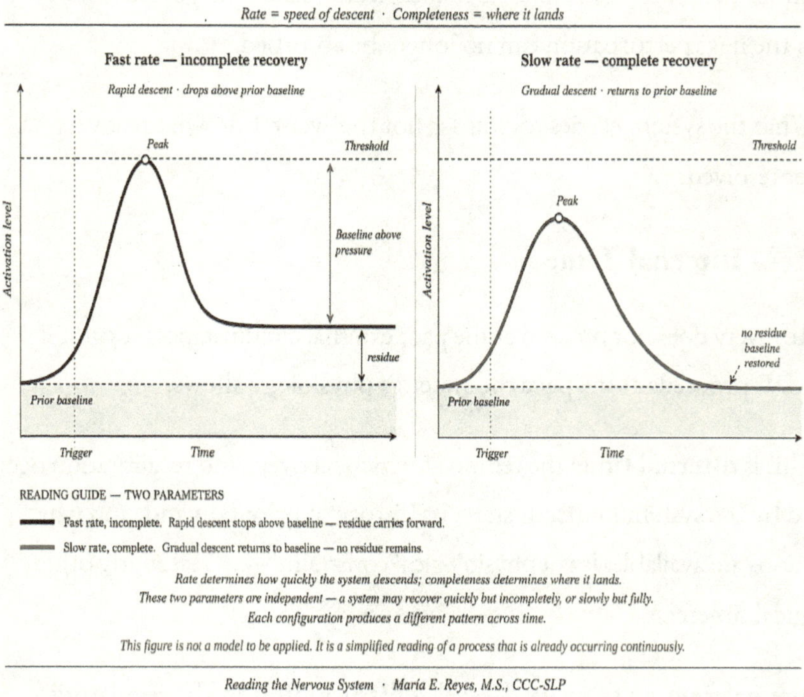

A system may descend at a moderate pace and stop short of its prior baseline. A system may return fully, but slowly—the descent is complete, but not finished before the next demand arrives. A system may recover both slowly and incompletely—each event leaving residue, with residue accumulating faster than it can be cleared. Each configuration produces a different pattern across time.

Rate and completeness operate between events—in the intervals, transitions, and quiet periods often assumed to be neutral. These periods are not empty. They are where recovery either occurs or fails to occur.

A system that has not fully recovered may continue to function adequately, showing no visible sign of the incomplete return it carries. The cost remains until it becomes visible. As residue accumulates, margin narrows. At that point, the next perturbation can no longer be absorbed.

What the system carries forward is not the event, but what recovery has not yet resolved.

4.3 — Internal Time

Recovery does not proceed at the pace external circumstances suggest it should. It proceeds at the pace the system's physiology allows.

This is **internal time**: the tempo at which recovery and recalibration occur, shaped by the system's current state, its history of prior demand, and what conditions are available. It is a physiological constraint — as real as any other biological timescale.

Internal time is a framework-level construct informed by regulatory dynamics that can be measured in part—parasympathetic recovery rate, heart rate variability, and vagal efficiency—which vary systematically across individuals and states.

Internal time varies across nervous systems and across states within the same system. A modest activation arc arising from a low baseline, with wide margin, may return toward baseline within minutes under supportive conditions. A steeper arc arising from an already elevated baseline, with narrow margin, may require substantially longer—not because the system is failing, but because clearing a larger activation event from a more depleted state takes more time.

The same duration. Different physiological meaning.

The distinction between internal time and clock time has practical consequences. A ten-minute break assumes ten minutes is sufficient for meaningful return. For some systems, at some levels of accumulation, it is. For others, it is not.

A break ends. The schedule resumes. The system continues recovering.

Asking a system to complete recovery faster than its current state permits is not neutral. It is an additional demand.

Rest that arrives before the system is ready becomes another form of demand.

A person instructed to return to baseline before internal time has allowed it is being asked to perform readiness. The performance may be possible. The physiological return it is meant to represent may not have occurred.

At the experiential level, internal time is felt as the pace at which readiness returns — the break is over, but the body has not yet finished what it was doing.

The felt sense of having recovered is itself a prediction the system generates about its own internal physiological state—and that prediction cannot stabilize without the interoceptive evidence that would support it (Seth, 2013). The instruction to recover does not produce that evidence.

The appearance of recovery is not recovery.

Internal time also introduces a form of mismatch—the same structural misalignment described in Chapter 2 for sensory load, now operating at the level of recovery. When environments, schedules, and relational expectations are built around assumptions about how quickly organisms recover, and those assumptions do not match the internal time of the system they are applied to, recovery remains incomplete by default.

Not slow, but not finished.

A note on empirical grounding: The construct of internal time draws on converging lines of research. At the autonomic level, it is informed by research on parasympathetic recovery rate and heart rate variability (Thayer & Lane, 2000; Shaffer & Ginsberg, 2017), including in multisensory integration (Stevenson et al., 2016) and in the temporal structure of attention and engagement (Karalunas et al., 2014). First-person accounts converge on this pacing dimension through descriptions of difficulty initiating, interrupting, or transitioning between actions (Buckle et al., 2021; Rapaport et al., 2024). These lines of evidence are developed further in Part II.

4.4 — Residue

When recovery is incomplete—when the system does not return fully to baseline before the next demand arrives—the activation that was not cleared carries forward. It becomes part of the baseline the system brings to the next event.

This carried-forward activation is residue.

Residue is not pathology. It is the ordinary consequence of regulatory work that has not yet fully resolved.

Under sufficient demand, incomplete recovery may leave functional carryover into what comes next.

Under supportive conditions—sufficient recovery time, low subsequent demand, and restorative sleep—residue clears, and the system begins the next period close to its prior baseline. (Sleep as a primary clearance mechanism is examined further in Part II.) The same conditions operate in reverse. Internal physiological conditions already being carried generate concurrent demand from within, compressing margin and elevating the baseline before any external demand has arrived. A body carrying these states does not begin from a restored baseline. It begins from one already loaded.

Residue becomes significant when conditions for clearance are absent—when recovery time is insufficient, intervals are shorter than internal time requires, or accumulation consistently exceeds clearance. Each incomplete recovery leaves the baseline fractionally higher than the one before it. (Residue, as used throughout this book, names the within-cycle and within-week dynamic of carried-forward activation. A related dynamic, operating at a much longer timescale and assessed through different measures, is what the allostatic load literature describes as sustained regulatory burden, addressed in §4.5.)

The floor rises. This dynamic becomes visible when traced across cycles. Figure 12 traces recovery and residue across cycles.

Figure 12 — Recovery and Residue Across Cycles

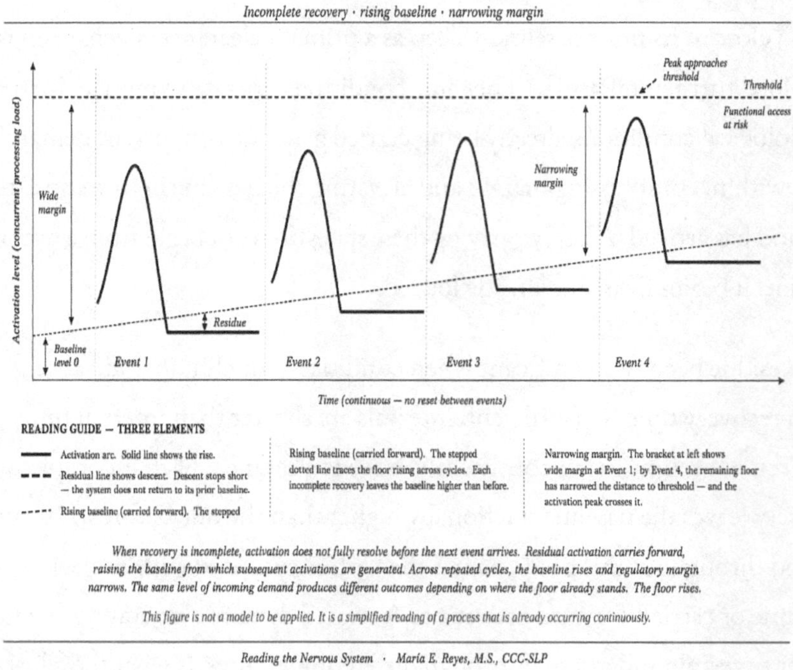

As the floor rises, margin narrows. Residue accumulates across time, even when each individual demand appears manageable.

A small event followed by a large response may be interpreted as disproportionate. One possible RNS reading is that it reflects a modest perturbation arriving in a system whose margin has already been narrowed by accumulated residue.

The system is responding to its state, not to the event alone.

4.5 — Accumulation Across Time

Within a day, accumulation appears as progressive narrowing of margin across successive demands. The system that managed the morning becomes less flexible by early afternoon. By evening, a stimulus that would have been absorbed earlier may produce a visible response.

The disruption appears in the evening. The accumulation began in the morning.

Across a week, accumulation appears as gradual elevation of the baseline floor. On Monday and Tuesday, the system may still be operating closer to a restored state. By Thursday and Friday, the floor has risen across multiple cycles of incomplete recovery. Less additional activation is needed to cross that threshold as the week progresses.

Friday afternoon is not a different system than Monday morning. It is the same system operating from a different baseline.

Residue across a week becomes legible as a pattern once it is named.

At the scale of months, a related dynamic may contribute to what the allostatic load literature describes as chronic regulatory burden—sustained elevation of baseline that reshapes autonomic tone, hormonal regulation, sleep architecture, and the parameters governing response and recovery. When sleep architecture is disrupted, the clearance mechanism that would otherwise reduce residue across the overnight cycle is compromised—and the system that begins the next period does so carrying load that was not cleared, not from a restored baseline.

McEwen and Stellar defined allostatic load as the cost of chronic exposure to fluctuating or heightened neural and neuroendocrine responses resulting from repeated challenge (McEwen & Stellar, 1993; McEwen, 2007). The mechanism this book calls residue—activation that persists past the point where it would otherwise resolve—shares a similar underlying logic: a system that has not been permitted to complete its return to baseline before the next demand arrives. It is not proposed here as identical to allostatic load, however: allostatic load is typically assessed through multisystem biomarker indices intended to capture cumulative physiological burden across longer timescales, while residue, as used in this book, describes a within-day and within-week pattern of possible carryover following incomplete recovery. The two are related in kind, not interchangeable in scale or in how they are measured. Figure 13 shows accumulation across longer timescales.

Figure 13 — Residue Accumulation Across Timescales

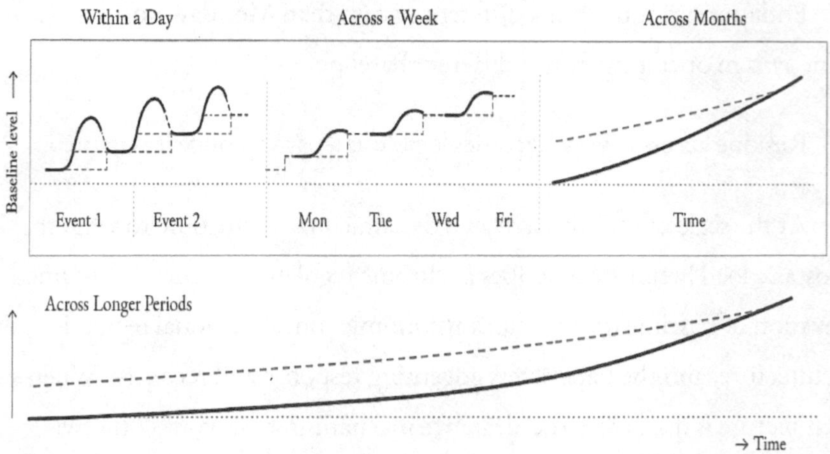

The same regulatory process appears differently depending on the timescale at which it is observed.

Within a day, accumulation is visible as incomplete return to baseline across successive events. Across a week, it appears as a stepwise elevation of the baseline. Across longer periods, it appears as a sustained upward shift in the system's operating level.

This figure is not a model to be applied. It is a simplified reading of a process that is already occurring continuously.

What follows traces what that rising floor produces. This accumulation does not proceed as a linear sequence, but as a reconfiguration of the entire system across cycles.

4.6 — Protective Stabilization

The rising floor §4.5 traced across days and weeks is also visible within a single arc—at the moment activation rises past what available margin can sustain.

Within a single activation arc, there are moments when activation rises, margin narrows, and recovery is not occurring quickly enough to restore functional access. At that point, the nervous system does not maintain normal function when the cost of doing so exceeds available resources. It reorganizes.

This transition becomes visible when the activation trajectory is traced across time. Recovery and protective stabilization diverge in Figure 14.

Figure 14 — Divergent Trajectories Under Load: Recovery and Protective Stabilization

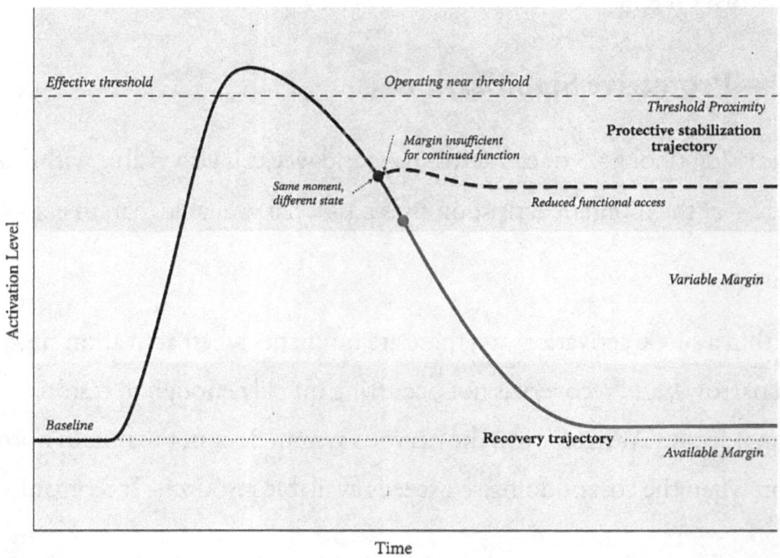

This figure is not a model to be applied. It is a simplified reading of a process that is already occurring continuously.

Reading the Nervous System • María E. Reyes, M.S., CCC-SLP

This figure is read across the two trajectories simultaneously—not as before and after, but as two different outcomes from the same activation event under different recovery conditions.

Degrees of freedom are reduced. Output is restricted. Input is limited. The range of activity being sustained narrows. Speech shortens or stops. Movement decreases. Social engagement withdraws. The attentional field contracts to the minimum required for basic orientation.

This is protective stabilization.

It is the nervous system doing what regulatory systems do when the cost of maintaining current function exceeds available resources: reducing function in order to preserve continuity and create the conditions under which recovery can proceed.

From the outside, protective stabilization is often misread. Withdrawal of speech, reduced engagement, and narrowed responsiveness may be interpreted as refusal or deliberate non-participation. The misreading matters because the response it generates is typically the opposite of what the system requires. Protective stabilization requires reduced demand, increased predictability, and time. Interpreting it as a behavioral problem tends to produce the opposite: increased demand, urgency, and consequence — each of which adds to load, narrows margin further, and prolongs the state.

Reading protective stabilization correctly requires looking not at the behavior but at what the system is managing beneath it.

What is intended as support becomes additional demand.

The instruction to re-engage arrives as additional load into a system that is already reorganizing to survive the current moment.

Protective stabilization diminishes as the conditions that generated it change. As activation descends, margin widens. As demand decreases, recovery becomes possible.

It does not resolve in response to demand. It resolves when demand is reduced.

4.7 — What Recovery Requires

Recovery is an active process that requires specific conditions.

Reduced concurrent demand is the first. Recovery proceeds less efficiently when the system is still managing high levels of unsuppressed signal. The physiological work of clearing activation draws on the same resources required for ongoing demand management. When both occur at once, recovery slows and completeness decreases. The system cannot clear what it is still required to carry.

Predictability is the second. Recovery is supported by environments that do not require continuous prediction updating—stable, familiar contexts low in sensory and social unpredictability. When the recovery environment itself is unpredictable, conditions associated with sustained apprehension and vigilance may also persist (Mineka & Kihlstrom, 1978; Grillon, 2008). The system remains partially engaged.

Time is the third. Not clock time, but internal time. When recovery intervals are shorter than the system requires, incomplete recovery becomes routine rather than occasional. The interval ends. The system is still returning.

Relational safety is the fourth. This is the presence of another nervous system whose state reduces rather than adds to load. A calm, predictable, non-demanding presence may reduce vigilance cost and support recovery; an urgent or unpredictable presence may have the opposite effect. Figure 15 names what recovery requires.

Figure 15 — Conditions Supporting Recovery Completion

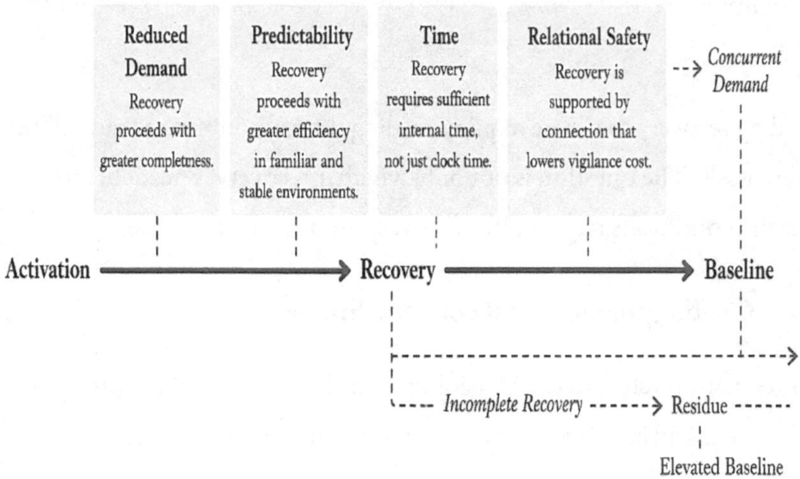

Recovery depends on specific conditions. When concurrent demand remains high, environments are unpredictable, time is insufficient, or relational context increases load, recovery slows or remains incomplete.

This figure is not a model to be applied. It is a simplified reading of a process that is already occurring continuously.

Reading the Nervous System • María E. Reyes, M.S., CCC-SLP

Recovery is not only individual. It is relational.

Within this framework, these conditions describe what the process of recovery may require in order to leave the system prepared for the next demand. When they are absent, recovery may remain incomplete.

This means recovery is not primarily an internal achievement. It is a condition-dependent process. Whether the interval becomes recovery depends on what the environment provides during it. The sensory, social, cognitive, and temporal demands of the environment continue shaping the system even after the original task has ended. A quiet room may not be sufficient if uncertainty remains high. Time alone may not restore what ongoing demand continues to consume.

Reading recovery therefore requires reading the environment as carefully as the system itself. The question is not only whether recovery is needed, but whether the conditions required for recovery are actually available.

4.8 — Co-Regulation as Recovery Support

Co-regulation refers to the physiological influence of one nervous system's state on another and its role as an active condition supporting recovery.

It is not primarily reassurance or problem-solving, though these may accompany it. It is the registration of another system's regulatory state—often below conscious awareness—as input that modulates one's own.

A nervous system in a calm, stable state provides signals—through vocal prosody, respiratory rhythm, facial tone, postural organization, and interactional timing—that are read as information about safety. When these signals indicate stability and low threat, vigilance decreases. Autonomic tone shifts toward the parasympathetic range that supports recovery.

Quiet reduces demand. A regulated presence reduces vigilance cost through its own physiological signaling.

Research on interpersonal physiological coupling shows that the physiological states of individuals in close proximity become coordinated over time—that the regulatory state of one person influences measurable autonomic parameters in another through the rhythmic patterning of shared interaction

(Feldman, 2007; Feldman et al., 2011; Coan & Sbarra, 2015).

Co-regulation is not unidirectional. A person carrying high activation or significant residue cannot provide the same co-regulatory input as a system that is well recovered. The state of the person attempting to support recovery becomes part of the recovery environment itself.

The signals that support co-regulation are not processed identically across nervous systems. For some systems, these relational cues are low-cost and stabilizing. For others, they may carry higher processing demand. This does not make co-regulation unavailable. It means the channels through which it is offered most effectively differ. The conditions under which co-regulation may increase rather than reduce concurrent demand are examined in Part II.

A note on what this account does and does not require: the co-regulation described here depends on the bidirectional relationship between autonomic state and social behavior, and on the role of parasympathetic tone in supporting recovery.
These claims are supported in the autonomic neuroscience literature (Thayer & Lane, 2000; Feldman, 2007). They do not depend on the specific mechanistic proposals of polyvagal theory—the phylogenetic hierarchy, the anatomical specificity of ventral versus dorsal vagal circuits, or the functional role of the unmyelinated dorsal vagal pathway—which remain contested (Grossman & Taylor, 2007). A practitioner who finds those specific claims unconvincing can still accept the account of co-regulation offered here on the basis of the broader literature. The polyvagal vocabulary is used in this book for descriptive utility, not as a commitment to its contested mechanistic proposals.

4.9 — What Accumulated Residue Looks Like

Accumulated residue is not random variability. It follows a logic: variability of access, delayed cost, and the structural consequence of environments calibrated to different recovery timescales.

Accumulated residue is expressed as patterns that may appear inconsistent from the outside but follow clear internal logic.

These patterns extend the mismatch principle. When environments are calibrated around recovery timescales the system cannot meet, residue accumulation is not an individual failure. It is a structural consequence of that calibration mismatch.

Within the framework, accumulated residue is modeled as elevating the baseline from which each new demand is met, narrowing available margin and shortening the modeled distance to threshold. Ordinary perturbations may therefore become more likely to produce responses that exceed what the event alone would appear to warrant. At the level of experience, this may register as the sense that everything is harder than it should be, that nothing feels fully restored, and that effort is required where ease was available before.

This becomes visible as the student who manages Monday, Tuesday, and Wednesday, and then cannot manage Thursday.

Capacity fluctuates—higher after adequate recovery, lower after insufficient recovery, and markedly reduced when accumulation outpaces clearance. The same task may be easy on some days and effortful on others.

What appears as inconsistency of ability is often variability of access.

Accumulated residue also alters what perturbations can be absorbed.

Small unexpected events generate larger responses. Transitions require more preparation. Social cues demand more processing.

Narrowed margin is encountering ordinary perturbations with the same activation slope it has always had.

Residue is also expressed as delayed cost. Visible difficulty appears at a point in time that follows the accumulation of load, not at the moment load first exceeded available margin. The proximate trigger is not the full cause. It is the first event that arrives after margin has already narrowed.

Reading residue requires attention to the arc of demand and recovery across the relevant timescale: when adequate recovery last occurred, what the system has been carrying since, and what intervals were available for clearance.

Environments carry implicit assumptions about recovery timescales: how long activation takes to clear, how quickly internal time can restore baseline, and how much accumulation is expected before the next demand arrives. For systems whose recovery operates on different timescales, those assumptions embed a structural disadvantage that has nothing to do with capacity and everything to do with calibration.

RNS organizes this changing field through four primary parameters: concurrent demand, activation slope, regulatory margin, and recovery. Considered together, these parameters also make visible additional properties of the field, including threshold proximity, internal time, and residue. Neither the parameters nor these additional properties are proposed as sequential stages or as directly measurable internal states. They are observational vantage points for examining how present demands, changes in organization, available capacity, recovery, and prior state may interact across time.

Chapter 5 — The Field in Motion

It is early morning. The same building. The same corridor. The same day beginning for both of them.

From the outside, nothing distinguishes one morning from the other. From within, they are already different.

Two nervous systems enter the same environment: the same fluorescent hum, the same ambient sound, the same visual complexity of a floor settling into its morning rhythm. Despite appearances, they are not beginning the same morning. The environment generates different levels of concurrent demand in each of them.

The same perturbation produces activation arcs of different steepness depending on the system receiving it. Recovery proceeds at its own pace. When that pace is insufficient, what was not cleared carries forward. Load becomes baseline. Baseline shapes what comes next.

Four parameters. Not separate in experience, but continuous in operation — four vantage points on a single regulatory field. Concurrent demand describes the starting state. Activation slope describes how quickly activation rises. Regulatory margin describes what room remains.

Recovery describes what the next cycle inherits.

This chapter reads them together. The scene that follows is an ordinary working day: an ordinary meeting, an ordinary afternoon.

The sequence that follows is an illustrative reconstruction using the framework's parameters; the internal states described are modeled for explanatory purposes rather than directly measured.

What differs is not the day. It is what each nervous system carries through it—and what each carries when it ends.

5.1 — Morning: The Baseline Each System Brings

By the time both people reach their desks, the morning has already been doing work.

For the first person, the commute was managed at low cost. The building environment is processed on entry and largely suppressed within minutes. Concurrent demand is moderate. The baseline brought into the first task is close to restored. Margin is wide.

For the second person, the commute carried more cost than it appeared to. The sensory variability of public transit—unpredictable sounds, proximity, continuous social monitoring—generated a succession of small activation events. The building environment has not fully suppressed. Concurrent demand is higher. Anticipatory activation around the first task rises from a baseline that is already elevated. Margin is narrower—though not yet at a level that produces visible disruption.

Neither person is in difficulty.

Both are at their desks, beginning their work, appearing fully functional.

The day has not yet made its first significant demand. The systems are already different.

The difference is not in attention, effort, or motivation. It is in the rate at which the environment's signals resolve into background—a function of processing architecture, not of the day.

5.2 — The First Perturbation

At nine forty-seven, an email arrives for both of them. A meeting scheduled for next Thursday has been moved to this afternoon at two o'clock. Preparation that had been distributed across the week must now be compressed into four hours.

Prediction fails for both. The same perturbation arrives in both systems at the same moment.

For the first person, the perturbation raises activation toward threshold. The slope rises, reaches a modest peak within a minute or two, and begins its descent. Margin is wide enough that the peak remains within functional range. By ten o'clock, the system has reorganized around the new timeline. Residue is small.

For the second person, the same perturbation crosses threshold into a system already carrying narrower margin. The slope is steeper. Activation rises more quickly and carries greater momentum. Settling begins — but from a higher point, in a system whose margin has already narrowed and whose recovery moves more slowly. Residue is larger. The margin available for the next demand is meaningfully reduced.

By ten o'clock, both people are preparing for the afternoon meeting.

Both appear functional.

5.3 — The Accumulation Before Two O'Clock

For the first person, the three-hour preparation period is demanding but manageable. Activation remains elevated but within available margin.

Language remains fluid. Planning remains flexible. Baseline rises only fractionally, and margin remains meaningfully intact. By one fifty-five, this person is ready.

For the second person, the same three hours carry higher cost. Small interruptions and variations—peripheral movement, minor sounds, notifications—arrive in a system that has been shifting all morning.

The floor has been rising.

Each event is minor. Together, they are continuous. Activation does not fully settle between them.

Incomplete recovery from each small event is elevating baseline and narrowing margin, increasing threshold proximity before the meeting's first demand arrives.

The preparation is completed. The output remains adequate. Nothing in the work itself marks the difference.

By one fifty-five, this person also appears ready. The margin entering the meeting is substantially narrower—not because the preparation was harder, but because the field in which it occurred carried additional concurrent demand.

Two people enter the same meeting at two o'clock. One with more available margin. One with less.

5.4 — The Meeting

The meeting begins. Both people are attentive and engaged. For the first person, activation remains well within available margin. For the second, it meets a margin that has already narrowed.

Twenty minutes in, a challenge appears. Someone raises an objection to the central argument—a substantive reframing that requires the prepared position to be reconsidered in real time.

For the first person, activation rises within a system that still has sufficient margin to contain it. Language remains fluid. Flexible thinking remains available.

In the second, the same objection arrives in a system whose margin has been narrowing since morning. Activation rises and carries past the range within which full functional access is available at low cost.

The key difference is not loss of function, but loss of ease.

Language remains available, but sentences require greater effort.

Elaboration becomes costly. Response is still possible. It is no longer inexpensive. What the meeting objection produced was not a response to that objection alone. It was the interaction between that demand and the state the system had been carrying since morning. From the outside, the response is adequate.

Analytic Pause — What the Day Built

Concurrent demand shaped the margin each person brought into the meeting. The meeting did not determine the starting state.

The commute elevated baseline concurrent demand. Each morning interruption narrowed margin through incomplete recovery. The steeper slope at nine forty-seven contributed to residue size. By two o'clock, those three mechanics had already shaped the conditions for the outcome.

The day did.

5.5 — After the Meeting

The meeting ends at three fifteen.

For the first person, activation begins its return—a gradual, uneven descent toward something approaching the morning baseline. Recovery proceeds within available margin. Under adequate conditions, baseline will be close to restored by the next morning.

For the second person, the return begins from a higher point. Margin is narrower. Recovery proceeds at the slower tempo that has been operating throughout the day. Residue is larger.

The difference is not only in how much each system carries, but in how recovery itself operates: its rate and completeness. Both parameters are operating under the weight of the day's accumulation.

The same meeting. The same objections. The same outputs. What they carry forward is not the same. Figure 16 traces this across a single day.

Figure 16 — Divergent Regulatory Trajectories Across a Single Day

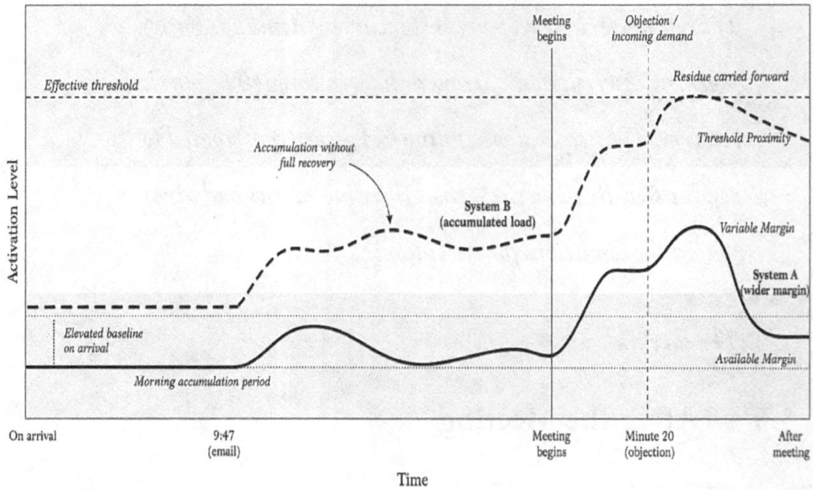

Two systems moving through the same day. Differences in concurrent demand and recovery across the morning produce different baseline conditions entering the same meeting. The same perturbation arrives in systems with different available margin. Divergence reflects what each system carries into the event.

This figure is not a model to be applied. It is a simplified reading of a process that is already occurring continuously.

Reading the Nervous System • María E. Reyes, M.S., CCC-SLP

The figure shows how incomplete recovery alters the starting conditions of each subsequent cycle. The residue each brings into tomorrow is different. The margin each will have available for the next demand is different. Across a week of similar days, the floor rises. At this longer timescale, this pattern shares a similar underlying logic with what the allostatic-load literature describes as repeated adaptation leaving a cumulative physiological burden (McEwen & Stellar, 1993; McEwen, 2007), though the two are not treated here as identical (see §4.5).

5.6 — Reading the Field

The day traced here is not a special case. It is an ordinary day read at the level at which it operates.

Following the four parameters together across time makes their interaction more legible. This is what it means, within RNS, to begin with the conditions surrounding behavior rather than treating behavior itself as a direct readout of internal state.

Behavior is often the last element to change.

It is the visible output of a field that has been in motion since the day began—shaped by what the system carried into the environment, what the environment required of it, how steeply activation rose when perturbation arrived, and how completely the system returned between events.

Accurate reading begins before visible disruption.

It begins with baseline. It follows concurrent demand as it accumulates. It traces the slope when prediction fails. It examines what recovery the intervals permitted and what residue they left behind. It does not wait for the surface to break in order to ask what is happening. By the time behavior changes, the accumulation that produced it has often been underway for hours.

The relationship between accumulated load and observable response is not linear. A small additional demand arrives in a system whose margin is already narrow — and what determines the response is not the demand alone, but what it encounters.

This is the practice: a way of reading that precedes the visible response.

A level of attention that follows physiology across time, before the threshold is reached.

The clinician, educator, or practitioner who has internalized this practice is not waiting for disruption to begin inquiry. They are already reading: the baseline the system brought to the morning, the demands the setting has been generating, the recovery the intervals have permitted, what has been carried forward.

Reading the field and responding within it are distinct practices.

What Part I builds is the first.

5.7 — Reading the Parameters in the Body

What the system can do at any moment depends on how much it is already carrying, how quickly activation is changing, how much range remains available, and how fully prior activation has resolved. These are not independent. They are four vantage points on a single regulatory field.

The parameters described in Part I may be reflected in recurring patterns of behavior, timing, effort, and bodily organization across contexts.

None of these observations identifies a parameter directly. The table below offers possible observational correlates that may become meaningful when considered across time, context, self-report, and competing explanations.

Table 2 — Reading the Parameters in the Body

Parameter	Definition	Possible observable correlates	Early changes	Across time
Concurrent demand	The total processing load the system is already carrying when a new demand arrives.	*Slower initiation; brief pauses before responding; gaze shifts or drops; reduced coordination; incomplete starts*	Momentary hesitation; need to pause before answering; slight disruption in sequence; increased reliance on repetition.	Pauses lengthen; actions become increasingly incomplete; reduced ability to hold multiple elements; previously manageable tasks fragment under accumulated load.
Activation slope	The rate at which the system moves from baseline toward higher activation when demand changes.	*Rapid changes in breathing (shallow or held); sudden muscle tension; quick shifts in voice; abrupt changes in movement; narrowing of attention.*	Small, rapid shifts in tone or posture; immediate tightening; early loss of peripheral awareness; urgency in responses.	Repeated rapid escalations in similar contexts; minimal transition between calm and overwhelmed; faster escalation across the day.
Regulatory margin	The available range within which activation can rise without exceeding what the system can sustain under its current conditions.	*Visible effort to stay organized; body held in tension; reduced spontaneity; limited tolerance for variation; careful, controlled engagement.*	Subtle rigidity; decreased fluidity of movement; effortful participation; reduced flexibility in shifting.	Increasing need for control; reduced tolerance for variation; more conditions required to sustain engagement; narrowing range of participation.
Recovery	The extent to which the system returns toward baseline after activation.	*Slowed return toward baseline; residual fatigue; reduced facial or vocal variability; withdrawal; decreased readiness to re-engage.*	Incomplete release of tension; continued shallow breathing; low energy after brief activity; difficulty restarting.	Baseline remains elevated; short breaks no longer restore readiness; longer recovery required; fatigue appears earlier across the day.

Also consider, before reading these signs as evidence of the parameter above: fatigue, unaddressed pain, an unrelated medical condition, task misunderstanding, valid disagreement, or motivation.

These parameters operate as coordinated changes within a single regulatory field. As concurrent demand increases and activation rises more rapidly, regulatory margin narrows. When recovery remains incomplete, baseline carries forward across time, and subsequent demands are encountered under altered conditions.

What changes is not the external demand, but the state that receives it.

That state is shaped not only by what the environment has been generating, but by what the body itself is carrying internally. These internal physiological conditions alter the readings across all four parameters before any external demand arrives.

Because the parameters operate within a single field, altering the conditions of one shifts the configuration of the whole—but not in a linear or predictable way. What changes is not a single mechanism, but the conditions under which the system organizes across time. This is one system, continuous across moments, shaped by what it carries and what it has been able to resolve.

. . .

Part I has described a shared regulatory architecture. Every nervous system carries demand into each moment, mobilizes in response to changing conditions, operates within an available regulatory range, and recovers across time. These are not features of any particular population.

Part II turns to autistic experience. It asks how this shared architecture may be organized across autistic individuals and contexts, and how documented differences in sensory processing, interoception, autonomic regulation, prediction, and recovery may interact with the regulatory parameters developed in Part I. The framework does not assume a characteristic autistic configuration of parameter values. Rather, it uses the same parameters to examine how individual processing patterns, current conditions, environmental demands, and recovery history may shape regulatory organization across time.

PART II — Autistic Nervous System Organization

Part II turns to autistic experience through the same regulatory architecture described in Part I. It asks how this shared architecture may be organized across autistic individuals and contexts, and how documented differences in sensory processing, interoception, autonomic regulation, prediction, and recovery may interact with the regulatory parameters developed in Part I.

The framework does not assume a characteristic autistic configuration of parameter values. Rather, it uses the same parameters to examine how individual processing patterns, current conditions, environmental demands, and recovery history may shape regulatory organization across time.

Recurring tendencies documented in autistic populations can inform this reading, but they do not define an individual nervous system. Parameter values cannot be inferred from diagnosis. They must be considered across person, context, time, and history.

The question is not what the system is, but how it has been required to organize.

When patterns that may reflect high concurrent demand or rapid activation are read only as noncompliance, avoidance, or dysregulation, the response may target output rather than the conditions generating it.

Table 3. Autistic Research Findings and Their Proposed Mapping to the Regulatory Framework

Part I construct	Evidence-informed observations in autistic samples	Framework interpretation
Concurrent demand	Sensory responsivity, arousal, and habituation findings are heterogeneous across studies and individuals; some autistic subgroups show persistent or elevated sensory-arousal responses, while others do not.	Under conditions in which sensory or other signals remain active longer, those channels may contribute to higher concurrent demand. This is a framework inference from partial evidence, not a demonstrated group-level value of the construct.
Activation slope	Autonomic reactivity findings are mixed and strongly context-dependent; current synthesis does not establish a generally steeper autonomic response in autistic people.	Particular individuals or contexts may show faster or larger activation trajectories, but a characteristically steeper activation slope should not be attributed to autism as a general property.
Recovery	Autonomic studies, including HRV/RSA work, provide evidence of group differences relevant to parasympathetic regulation and recovery, alongside substantial individual and methodological variation.	These findings may indicate differences in autonomic components of recovery for some autistic individuals or contexts; they do not directly establish the framework's full multisystem recovery parameter.
Internal time	No direct empirical literature identified here validates the framework's within-cycle construct of internal time as generally longer in autistic people. Burnout and transition/inertia literatures address related but non-equivalent timescales.	Longer integration or recalibration time is a framework-level hypothesis informed by converging observations, not an established autistic characteristic value.

Note on construct status. Regulatory margin (defined in Table 1) is not listed separately because, within the framework, narrowing margin would be expected when concurrent demand is elevated and recovery remains incomplete. Whether and how this occurs in any individual must be established through observation rather than inferred from diagnosis. Internal time, while derived from the primary parameters, is included because it provides a distinct observational dimension for considering how continuity, transition, and closure are experienced within the system.

The mappings above are not direct measurements of framework parameters in autistic populations. They show how heterogeneous empirical findings may be interpreted through the framework. Individual parameter values must be inferred cautiously from patterns across contexts and time rather than assumed from diagnosis or group-level findings.

This asymmetry can make mismatch structural rather than occasional. Many environments are calibrated around regulatory assumptions that may not match the processing conditions of the autistic person within them.

As physiological conditions accumulate across time, the organization of the nervous system may become harder to read from any single event.

Responses begin reflecting not only the immediate environment, but also unresolved activation, altered thresholds, anticipatory prediction, and recovery history. The same outward behavior may therefore emerge from very different underlying physiological organizations.

Chapter 6 — Autistic Interoception: Another Way of Inhabiting the World

Opening — The Architecture in Motion

The sequence that follows is an illustrative RNS reconstruction rather than a direct account of measurable internal state.

The night was adequate but not fully restorative. Internal time required more than the available hours. The person reaches the kitchen carrying a floor that is fractionally elevated — the physiological residue of sleep that did not complete the clearance the system required. Below the threshold of awareness, invisible to anyone observing.

The commute involves unpredictable sound, shifting proximity, and continuous social monitoring. The fluorescent hum of the train car has not resolved into background by the time the destination is reached. It remains active, occupying resources as the next segment of the day begins. Concurrent demand is already elevated before the first task appears.

The margin entering that first demand is narrower. This is what ordinary transit costs this architecture.

An unexpected message requires a response before the first task has begun. Prediction fails. Activation is recruited quickly. By the time the response is sent, the arc has already progressed, leaving residue that will not resolve before the next event arrives. Recovery has not completed, but the sequence continues.

Mid-morning: two hours of adequate work. The environment is not quiet. Background movement, a conversation two desks over, and three unanticipated questions each cross threshold. Each leaves a fractional remainder. The margin available entering the eleven o'clock meeting is narrower than it was at nine. None of this is visible.

The work is intact. The baseline is rising.

What is not aligned is the timing. Activation progresses. Access follows later. Recovery extends beyond the interval available before the next demand arrives. The system continues forward while previous activity remains physiologically active within it.

What is not aligning is the timing across cycles — the rhythm through which activation, integration, and recovery would otherwise reach completion.

What organizes here is not a response to an event, but the way signals are sustained, transformed, or left unresolved within the system.

This is an ordinary Tuesday. Not a crisis. A description of cumulative dynamics under conditions that do not allow full resolution between events.

Coherence, continuity, and closure vary accordingly. Experience does not fully integrate within a given state. What has not resolved carries forward into what follows, and the next sequence begins from that condition rather than from a returned baseline.

6.1 — Autonomic Organization in the Autistic Body

Two people arrive at the same building within minutes of each other. The same entrance. The same corridor. The same fluorescent lighting cycling at its characteristic frequency. The same ambient sound from the ventilation system. The same cluster of colleagues near the coffee station speaking at normal conversational volume. The same smell. The same visual complexity of a busy open-plan floor settling into its morning rhythm.

From the outside, both have arrived. The environment is identical.

The morning has not yet made any demand.

For the first person, the fluorescent hum registers briefly on entry and is suppressed within seconds — marked as non-informative and removed from active processing. The ventilation sound never reaches awareness. Peripheral movement from colleagues is noted and released. The visual complexity of the floor is scanned once and organized into a stable perceptual pattern that will not require re-examination unless something within it changes. By the time this person reaches their desk, the environment has largely been down-weighted. What remains in active processing is the task they are moving toward.

For the second person, the fluorescent hum remains present — not as a complaint and not consciously attended to, but as a signal that has not yet been suppressed and continues to occupy processing resources. The ventilation sound resolves more slowly into background. Peripheral movement registers each time it shifts. The visual field does not stabilize as quickly — movement across it continues requiring attention rather than receding into unattended background. By the time this person reaches their desk, the environment has not been resolved. The body has already begun adjusting to what the environment has been generating.

The shift is underway before conscious access has organized around it.

Neither person is doing anything wrong.

What differs is the threshold at which signals are suppressed, the speed with which repeated input is down-weighted, and the amount of processing the environment continues to require. The sections that follow trace what produces that difference.

Autonomic regulation and interoceptive access do not always unfold at the same pace. One process organizes changes in the body across systems; the other reflects when those changes become consciously accessible as experience.

The underlying architecture is shared across nervous systems. What differs is the timing and relationship between these processes as they unfold.

In many autistic profiles, the body may already be moving into activation before the person can clearly recognize or interpret what is happening internally. Within this proposed sequence, activation may begin before conscious interoceptive access organizes around it.

This separation becomes visible under load. As activation crosses threshold, it may continue to rise quickly, before interoceptive access organizes around it. By the time experience becomes available from within, the system may already be operating at a different point along the arc.

The system may already have shifted before the felt sense becomes available.

Measures of autonomic activity can index aspects of this organization — the shape of activation, the pace of return, and the available room before the next demand. They do not define it. The relevant question remains local to the system in its current state: what is still active, what has not yet resolved, and how much return toward baseline the available interval permits.

6.2 — Fine-Threshold Calibration and System Consequence

Within this framework, the timing relationships described in §6.1 —

between signal persistence, autonomic activation, and conscious interoceptive access — are proposed to have a specific organizational consequence: a system that resolves signals more slowly may also carry more of each cycle into the next. As Table 3 illustrates, autistic profiles vary widely in how these timing relationships present; the organization described below is illustrative of one common configuration, not a universal calibration.

Environmental and bodily signals that other systems down-weight more quickly continue to occupy resources. The system remains engaged with information that has not yet resolved. Across a day without sufficient intervals for recovery, what begins as fractional residue becomes a rising baseline, shaping the conditions under which each subsequent demand is met.

This reflects a system that registers change at a finer threshold. More input persists in processing rather than being rapidly suppressed. Variation is retained — including internal signals — beyond the point at which other systems have already stopped actively processing them.

This does not change how activation unfolds — it changes where the process begins and how much information continues drawing on processing resources as activation unfolds. More signals enter the arc, increasing what the system must differentiate within the same interval. Activation is recruited more frequently. Recovery must account for a greater accumulation of unresolved signal. The interval required for sufficient recovery often extends beyond what the environment provides.

Each cycle begins with less available capacity than the one before.

When concurrent demand is reduced and sufficient time for descent is available, this same organization produces a different consequence. Retained variation does not accumulate as unresolved load. It remains available as detail, allowing differences that would otherwise contribute to rising baseline to be integrated within the same cycle.

Sensitivity increases cost when time is insufficient; when time is available, it supports resolution.

What follows moves from threshold calibration to how prediction itself operates under this organization.

6.3 — Interoceptive Prediction in the Autistic Body

Prediction shapes which signals are resolved and which remain active as the system moves forward. What differs is how much the system relies on immediate input rather than prior patterns — and how quickly uncertainty can settle within the time available.

Under this organization, prediction relies less on rapid compression into prior patterns and more on the detail of ongoing input. Immediate signals retain greater weight relative to previously established expectations. As a result, resolution depends more on processing what is immediately present than on quickly matching experience against prior patterns. Some accounts of autistic perception describe this as reduced reliance on prior expectations, producing a world experienced with sharper, less compressed signals (Pellicano & Burr, 2012). These accounts remain under active development; what the framework draws on is the organizational description rather than any specific mechanistic claim.

When integration depends more heavily on the detail of what is immediately present, ambiguous input does not resolve quickly. Multiple possibilities persist simultaneously, each continuing to draw on processing resources. Processing extends across time rather than closing within the moment — more remains open as the next demand arrives, extending the activation arc and narrowing available margin.

This becomes especially visible in environments where cues are variable, implicit, or rapidly changing — where resolution would depend on rapid compression into prior patterns. When that compression does not occur within the available interval, ambiguity persists and each cycle begins with reduced capacity to differentiate what follows.

Under these conditions, activation may already be progressing before it becomes available to conscious access. The delay between what is happening in the body and what becomes consciously available may not be a separate feature — within this framework, it can be read as reflecting how much the system is still carrying and how quickly activation is moving. At the level of experience, what is being felt reflects a later point in a process already underway. Figure 17 maps activation and interoceptive access across time.

Figure 17 — Activation and Interoceptive Access Across Time

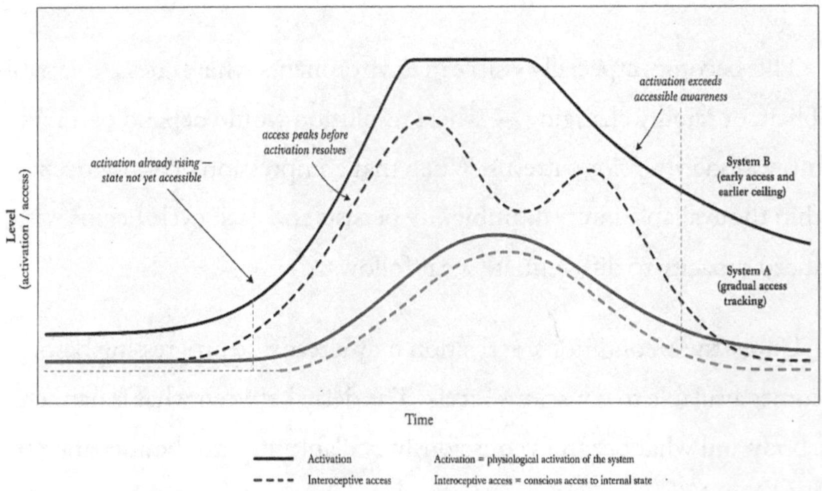

Activation and interoceptive access do not unfold simultaneously. The system may already be in motion before the state becomes accessible. In System B (early access), interoceptive access emerges early but does not necessarily track increasing activation across time. In System A (gradual access), interoceptive access tracks activation more closely.
This figure is a simplified illustration of two hypothetical patterns, not a measurement of any individual's activation or interoceptive access.

Reading the Nervous System • María E. Reyes, M.S., CCC-SLP

Patterns consistent with this organization have been described in studies of predictive processing in autism (Lawson et al., 2014). Interoceptive difficulty has also been documented in autistic samples, although Brewer et al. (2016) found that these differences were associated with co-occurring alexithymia rather than autism itself.

What determines how long that processing extends is not only how signals are weighted, but the pace at which recalibration can occur.

6.4 — Autistic Internal Time

Internal time sets the pace at which recovery and recalibration can occur. It does not accelerate because urgency demands that it do so. What cannot resolve within that pace is carried into what follows.

In some autistic profiles, the interval available does not match the rhythm the system requires to complete its cycle, extending beyond what many environments provide between demands.

Integration requires staying inside one state long enough for it to complete, and activation needs time to settle before the next demand begins. When the next demand arrives before that process completes, the system proceeds without closure — what was active continues into what follows. Baseline rises, and regulatory margin narrows in advance of subsequent demands.

The system is still organizing what came before while already engaged with what comes next. Figure 18 shows accumulation across cycles.

Figure 18 — Internal Time and Accumulation Across Cycles

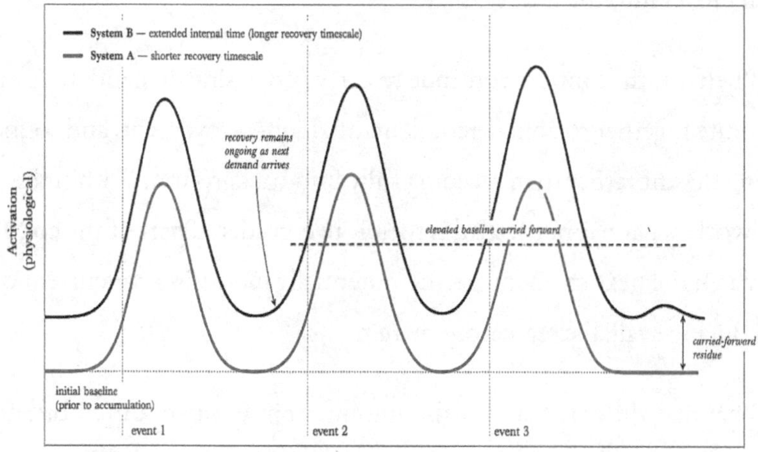

Same events. Same environment. Different recovery timescales. When recovery remains ongoing as the next demand arrives, activation carries forward and baseline rises across cycles. Values are schematic representations.

Reading the Nervous System • María E. Reyes, M.S., CCC-SLP

Under these conditions, the difference across the day is not capacity but carryover. Events that were manageable earlier occur from a different starting point later — not because the system has changed, but because recovery has not completed within the time available. When the interval between demands is shorter than the time required for resolution, activation does not clear before the next sequence begins, producing a gradual elevation of baseline and a progressive narrowing of regulatory margin across hours and days.

The same dynamic operates at the level of sensory input, where what enters the system does not resolve at a single rate.

6.5 — Sensory Organization and Regulatory Demand

The same pattern can be considered at the level of sensory input. Inputs do not necessarily recede in processing weight at the same rate across individuals or conditions. Within an RNS reading, input that settles rapidly into background for one person may remain more salient or demanding for another as the system moves into the next demand.

When multiple inputs continue to carry processing demand, they may contribute together to concurrent demand. Light, movement, and sound can remain relevant rather than receding fully into background. Within the framework, what has not settled is therefore considered part of the conditions into which the next moment arrives, potentially increasing concurrent demand and reducing available regulatory margin.

What may differ is not only the amount or intensity of input, but how readily elements of the environment recede into background. When this occurs more slowly, ordinary surroundings may remain an ongoing source of processing demand. RNS reads the resulting effort or fatigue not as proof of a particular sensory mechanism, but as potentially reflecting the cost of sustaining organization while multiple inputs remain relevant.

Sensory processing, in this framework, is not treated as separate from regulatory activity; sensory conditions are considered among the factors that can shape regulatory organization. The relevant cost may involve not only stimulus intensity but also the continuing processing required when input remains salient or unresolved.

Donna Williams described the sensory environment as overwhelming and difficult to filter, with multiple inputs remaining active at once (Williams, 1992).

Across environments organized around faster resolution, this difference may become structurally consequential. Within this framework, when sensory input remains active longer, the system may carry more into each subsequent moment — not because more is present, but because less has cleared. Patterns consistent with slower sensory habituation and sustained processing of input have been reported in studies of sensory responsivity in autism (Merchie et al., 2023; Gonçalves et al., 2023).

Paired Scenario — Sensory Load and Concurrent Demand

It is the same classroom. Morning. Both children have arrived at their desks. The room carries its ordinary conditions: fluorescent light overhead, ambient movement along the periphery, voices from the corridor, the low hum of equipment.

For the first child, the room resolves within a few minutes of entry. The overhead light recedes. The corridor sound diminishes in processing weight. What remains active is the task.

For the second child, the fluorescent hum remains present.

Movement at the edge of vision continues to engage orienting resources. Each sound from the corridor carries the same processing weight it did on entry. The room has not resolved into background. It is still arriving.

The teacher issues the first instruction. The second child receives it while also managing what has not yet settled.

Both children are seated. Both appear ready. The load entering the task is not the same.

6.6 — Recurring Interoceptive Orientations

What §6.1 through §6.5 have traced can be read within this framework as a recurring pattern at different levels: threshold, prediction, internal time, and sensory input converge on an organization in which activation, access, and recovery do not necessarily run at the same pace. Across these levels, unresolved activity may continue to organize what follows.

Chapter 7 traces what this organization means for the level at which thought is constructed and language becomes available.

Chapter 7 — Embodied Cognition: The Body as the Ground of Thought

Opening — Cognition in the Same Regulatory Field

A configuration is assembling. Not in words. In relation — the way several things are beginning to fit. There is no name for it yet. The name would come later, if the assembly were allowed to finish.

Someone speaks. The question is reasonable. An answer is expected.

The assembly stops. Its parts remain, but the relations among them have dissolved.

What was being built cannot be resumed. It must be rebuilt — which is not the same as continuing.

The understanding was almost there. The interruption did not arrive after. It arrived instead.

This reflects a mode of cognitive organization that depends on sustained continuity — interrupted before the conditions required for closure were present. What is being described here are the conditions under which thought can complete.

What is interrupted here is not the content of thought, but the conditions that were allowing it to organize. What disappears is not information, but access to the state in which that information was cohering. What was being sustained was a configuration of signals organized around a single process — and within the reading proposed here, interruption may disrupt that configuration rather than simply pause it.

What follows reads the conditions under which thought organizes — not the content of thought itself, nor the meanings it generates, which extend beyond the level of organization this framework addresses.

7.1 — Thinking From the Body

Cognitive processes unfold within an ongoing state of physiological organization (Varela, Thompson, & Rosch, 1991; Barsalou, 2008; Seth & Friston, 2016). What varies across individuals and contexts is how physiological organization interacts with what can be thought, remembered, or expressed — a feature of any nervous system, not of autistic cognition alone. In autistic experience, documented differences in sensory, interoceptive, predictive, and autonomic processing may make some of these state-dependent relationships especially important to examine, without implying a single autistic regulatory architecture.

What is available for thinking, language, and integration shifts more detectably as physiological conditions shift — not because capacity is unstable, but because the margin within which access can be sustained is narrower and more specific. A change in activation level, a shift in sensory load, or an incomplete recovery arc more immediately shapes what the system can organize. A higher degree of coupling between physiological state and cognitive availability makes the regulatory conditions of thought more readable.

In some autistic profiles, bodily signals serve as the primary ground from which thinking organizes — shaped by the body's current state, how integrated that state is, and the rhythm of activation within it. A shift in internal rhythm, a particular sensory experience, or a change in the system's global organization may initiate association, support the emergence of understanding, or facilitate creative movement. Thought may first emerge as organized experience and, under conditions that support the system's pace and integration requirements, become available to language.

7.2 — Sustained Focus, Internal Time, and the Cost of Interruption

The form cognition takes reflects what the system can sustain.

What is being sustained is not only access, but the rhythm through which thought can organize and complete.

In many autistic profiles, attention and thinking may organize around a single focus and remain there until the process has reached sufficient integration. This is a monofocal mode of processing — not a limitation of attentional breadth, but a form of cognitive organization that prioritizes depth and sustained continuity over speed and frequent switching (Murray, Lesser, & Lawson, 2005; Mottron et al., 2006).

When conditions support it, this mode can sustain forms of thinking that rapid switching often disrupts: engagement with a problem long enough for deeper structure to become visible, associative connections that emerge only after sustained continuity, and integration that approaches closure rather than ending at a partial pass.

Integration requires continuity: the system must stay inside one state long enough for experience to organize without fragmenting — holding a thread of processing until closure is reached before the next demand arrives. The duration reflects the depth of integration being sustained, not processing speed.

Autistic accounts consistently describe the cost of interruption. One autistic adult described it this way: "where I was isn't a position — it's a state that existed while everything was held together" (Yergeau, 2018).

When interruption arrives, what was being built does not pause; it ends, and must be rebuilt from the beginning. This captures both the cost of interruption and the loss of an integrated state that cannot be resumed without reconstruction.

In many of these cases, interruption takes the form of a premature transition — a shift imposed before the system has reached sufficient integration to complete the ongoing process.

Each premature transition — whether driven by external interruption or environmental pacing — leaves an unresolved process that continues to occupy processing resources and contributes to concurrent demand as the system engages the next task, reducing the margin available for what follows. In environments characterized by frequent interruptions — speed demands, multitasking requirements, or rapid transitions between tasks — this accumulation of incomplete processes compounds across the day.

The cost follows from a processing organization oriented toward depth encountering environments that require transition before integration can reach completion.

Enlarged multisensory temporal binding windows have been reported in autistic samples, particularly in audiovisual integration, although effects vary across tasks and stimulus types. These findings provide one example of how temporal organization may differ in autistic experience, but the psychophysical timescale measured by temporal binding paradigms is narrower than the broader regulatory construct of internal time used here.

Autistic inertia makes it visible from the inside: Buckle et al. (2021) found that participants described difficulty with transitions not as reluctance or motivation problems but as a gap between when the environment expects the shift and when the system is prepared to make it. Murray et al.'s (2005) monotropism account converges from a different angle— attention tends to organize around a single strong processing thread, with high intensity and resistance to redirection: a mode in which internal time is shaped by how deeply one process integrates rather than how many run in parallel.

Paired Scenario — Sustained Focus, Interruption, and Reconstruction Cost

It is a structured work period. Both students are engaged in the same assignment.

The assignment continues. Both students complete it. The cost of arriving at completion is not the same. For the second student, each interruption does not slow the work. It restarts it.

What is encountered here is not a difference in capacity, but a difference in temporal alignment: the pace at which the environment requires response does not coincide with the time required for integration to reach completion.

When integration must be sustained rather than rapidly completed, the forms of thought that become possible are those that hold structure across time — configuration rather than sequence.

Interruption does not only break continuity; it disrupts the rhythm required for integration to reach closure.

7.3 — Pattern-Based Organization and Preverbal Thought

Under these conditions — where sustained sequential formulation cannot be maintained — the system does not stop organizing. It organizes differently.

In many autistic accounts of thinking, and across some autistic profiles, thought may organize prior to language. When sequential formulation cannot be maintained across internal time, meaning does not assemble step by step but organizes around what can be held in continuity. Under these conditions, meaning may take shape as a felt whole — an internal scene, a spatial relation, a rhythm, a bodily adjustment, or a pattern emerging from the integration of perception and interoception. Meaning consolidates as a global structure stabilizes rather than through rapid sequential assembly toward a named conclusion.

Temple Grandin (1995) describes thinking as primarily image-based, with words arriving later in the process. What this account illustrates is not a deficit in verbal processing but a difference in primary medium of organization: meaning may assemble configurationally, as an integrated whole, before sequential verbal formulation begins.

In some profiles, the nervous system may be especially responsive to regularities that other systems treat as secondary to fragments or context. Attention to detail, sensitivity to subtle differences, and detection of inconsistency can support a form of pattern-based cognition in which thought advances as relations among elements stabilize. Language may arrive later; understanding may already be present before verbal expression has organized around it.

Bodily memory may participate in this configurational recognition, allowing similar patterns to be anticipated or completed without requiring explicit recall (Damasio, 2010; Fuchs, 2012). The internal scene is not exclusively visual. It may be lived as rhythm, tension, movement, spatial orientation, or a felt sense of fit — a whole apprehended before its parts have been separately named.

This mode of organization carries a particular sensitivity to interruption. When a configurational process is disrupted before the global structure has stabilized, the integration underway may not be directly recoverable — the system must reconstruct rather than resume. Available margin, recovery time, and concurrent demand determine what the next demand must be met from.

This dynamic may extend beyond cognition alone. Emotional and relational processes that organize as whole-field experiences may show similar sensitivity to interruption: an arc disrupted before sufficient organization may not simply pause. When such interruptions recur, the incomplete arc becomes part of the starting condition for what follows.

Environments that interrupt before closure — requiring verbal output before understanding has completed its organization, or interpreting silence as absence of processing rather than as ongoing processing — may systematically generate this cost.

Output remains possible under interruption. What changes is what the next demand must be met from — and what remains for the one after that.

7.4 — The Bodily Filter: A Clinical Reading of Cognitive Access

When the system is operating within adequate margin — when concurrent demand is low and recovery has been sufficient — constraints on access are reduced. Integration can proceed across multiple streams and language becomes more available, provided the environment does not generate sustained activation cost.

As margin narrows, the range of signals that can be integrated correspondingly contracts. Capacities that previously supported elaborate synthesis may now support only partial integration. This shift reflects not a loss of capacity, but a change in the conditions under which that capacity can be accessed.

This state-dependence extends across domains. Cognitive, emotional, and relational access follow the same regulatory conditions. Access is not domain-specific; it reflects the system's current organization. In autistic profiles where the conditions for access are more precisely specified, access tends to be more state-sensitive — because it requires more specific conditions to become available. Figure 19 maps the relation between regulatory margin and functional access.

Figure 19 — Regulatory Margin and Functional Access

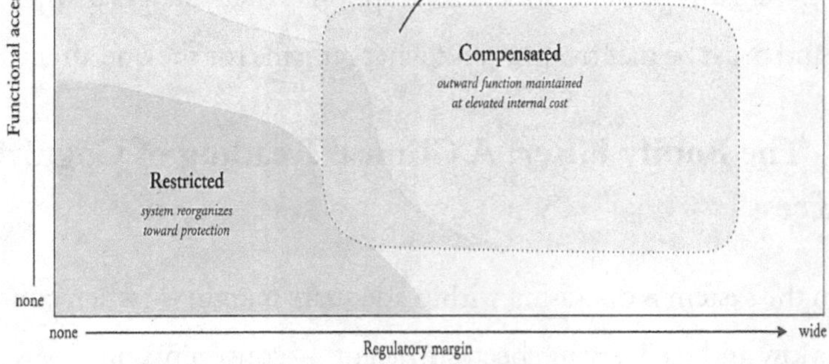

Positions within this field describe states, not categories — locations within a continuously shifting field shaped by current activation, recovery status, and concurrent demand. Compensated positions reflect maintained outward function despite narrowed effective margin and elevated internal cost. Values are schematic representations.
Reading the Nervous System • María E. Reyes, M.S., CCC-SLP

This figure maps functional access as a state-dependent position within the field rather than as a fixed trait. The horizontal axis represents regulatory margin — the available range before organized functioning degrades. The vertical axis represents functional access — the degree to which capacities such as language, emotional organization, executive functioning, and flexible engagement are currently available.

When margin is relatively wide, conditions may support broader functional access. As margin narrows, functional access may become increasingly constrained even while functioning remains outwardly organized and requires increasing internal effort, producing compensated access. Under fluctuating residue or incomplete recovery, access may become variable across time despite unchanged underlying capacity. When margin narrows beyond what organized functioning can sustain, access may become restricted and the system may reorganize toward protection.

The arrows indicate movement across the field as accumulation narrows margin and recovery restores it.

What appears as variability in attention, comprehension, or response is not inconsistency in the person, but variation in what the body is able to organize at that moment.

7.5 — Communication and Bodily Access

Communication unfolds from the system state that can sustain expression. This requires the system to hold internal continuity while keeping an expressive channel open; as margin narrows, lower-demand communication channels may remain more accessible than channels requiring greater simultaneous coordination.

The pattern is visible when a child who speaks fluently at home arrives at school and goes largely nonverbal by mid-morning—not because language has been lost, but because the conditions that support access have narrowed.

When margin is wide, multiple channels may be available at once: spoken language, elaborated response, relational attunement, eye contact, and the reciprocal flow of conversation. As margin narrows, channels reorganize according to relative cost. Spoken language—which requires the coordination of lexical retrieval, syntax, prosody, relational calibration, and sensory monitoring—may become less accessible. Written, gestural, or other lower-demand channels may remain available under these conditions. This reasoning is consistent with clinical models such as SCERTS, which incorporate multimodal communication supports rather than assuming that spoken language is the only available channel (Prizant et al., 2006).

Within this framework, this variability can be read primarily in relation to physiological state rather than assumed to reflect will, cooperation, or communicative intention. When environments assume that speech is continuously available, demand may increase at the point where access is already constrained, further reducing the conditions under which communication can occur.

Communication does not disappear when speech is absent. It reorganizes within the channels that remain accessible under the system's current conditions.

Closing — When the Body Sustains Thought

Cognitive organization, language, and communication unfold within the system's regulatory conditions.

What becomes available as thought, expression, and exchange reflects the system's current state — the margin available, the activation being carried, and the extent to which prior activity has been metabolized. As these shift, what can be organized shifts with them. Variability in access does not reflect inconsistency of capacity, but the changing conditions under which that capacity can be brought into form.

This is an architecture in which access is not given, but generated — shaped by the conditions the environment supports or constrains, and by the state the system can maintain within them.

Chapter 8 — Bodily States, Energy, and Regulation in Autism

Opening — The Cost That Accumulates

The day begins without difficulty. The environment is familiar. The tasks are known. Each demand, taken on its own, is manageable. There are small adjustments throughout the morning — a sound that requires filtering, a transition that takes longer than expected, a social exchange that requires sustained attention. None of these exceed the system's capacity. Each one is absorbed. The work continues. By early afternoon, nothing has changed in what is being asked. The same instructions. The same environment. The same expectations. But responses begin to shift. The time it takes to respond lengthens. Transitions require more effort. Minor changes that would have been absorbed in the morning now produce hesitation, or resistance, or withdrawal. No single event explains the change. What has changed is what the system is carrying.

What changes across the day is not the demand, but the baseline from which it is being met.

Chapters 6 and 7 read autistic parameter values within single cycles and across cognitive access. This chapter shifts scale: from what a single cycle costs to what accumulates when recovery cannot restore what expenditure removes.

The question follows directly: what accumulates when physiological expenditure consistently exceeds what recovery can restore?

What the system is carrying is not the memory of an event. It is signals that have not resolved — configurations still active, shaping how the next demand will be met.

The floor has been rising across days, not within a single moment.

What emerges from that rising floor are distinct bodily states — configurations through which accumulated load shapes regulation across time. The system's available range narrows through two pathways: acute overload and sustained exhaustion. When the exhaustion pathway is not recognized, a further condition emerges that the literature describes as autistic burnout.

The focus shifts here from single cycles to accumulation across cycles.

When expenditure and recovery remain in sufficient balance, the system remains available. When expenditure consistently exceeds recovery, accumulation unfolds across time. Baseline elevates, reducing the margin available for what follows.

What follows traces the consequences of that progression as a descriptive physiological account.

8.1 — Baseline State as a Physiological Condition

Baseline state is the physiological ground from which everything else becomes possible.

In many autistic profiles, this dimension is not directly visible. Behavioral organization may be maintained until collapse, obscuring the load the system has been carrying (Raymaker et al., 2020; Mantzalas et al., 2022). Regulation is often inferred from the behavioral surface: a person who is participating, complying, or producing appropriate output is assumed to be well regulated.

But the absence of visible difficulty is not evidence of low load. It indicates only that the system is managing its load within the margin currently available.

What remains hidden is the condition from which that organization is being produced. Baseline may be elevated. Concurrent demand may be high. Recovery may be incomplete. Behavior can remain organized while the system operates from reduced margin.

What emerges from those conditions is not only how the system regulates, but what becomes available.

Access is the clinical face of regulation: whether the system's cognitive, communicative, and relational capacities are currently available — whether the person can remain present, respond, and relate under the demands currently present.

To read regulation accurately is to read what the system can currently organize, sustain, and produce from the state it is in, rather than from the state it is expected to be in.

The relevant question becomes: from what state is expression being organized?

8.2 — Bodily States as Configurations of the Four Parameters

Bodily functioning becomes more intelligible when read through system states — dynamic configurations reflecting the relationship among concurrent demand, activation slope, regulatory margin, and recovery.

These states are not fixed categories or diagnostic labels, but positions within a continuous field.

Reading through states rather than behaviors keeps explanation at the level of producing conditions rather than produced output. The same nervous system may be in Organized Availability on Monday and in Exhaustion by Friday — not because its architecture has changed, but because its parameters have been operating under different conditions.

Organized Availability is the state in which margin is sufficient. Concurrent demand is managed within capacity. Baseline has returned toward a level that supports continued engagement. The system can produce, respond, and remain available without those functions consuming what remains.

Activation is the state in which demand is increasing and physiological response is being recruited. The slope rises. Mobilization increases. What matters is whether available margin can contain that rise and whether recovery can occur before the next demand arrives.

Overload is the state in which the activation arc exceeds available margin. Language narrows. Flexibility contracts. Choice reduces. Protection begins to organize.

Exhaustion differs from overload in both mechanism and timescale. Where overload reflects activation exceeding available margin in the present moment, exhaustion reflects a baseline that has been elevated across time.

These positions describe where the system is relative to its margin — not separate conditions, but locations within a continuously shifting field.

8.3 — Regulation as Dynamic Balance

Regulation is cyclical — movement through activation and return toward baseline across the day. When recovery is sufficiently complete, margin is restored toward its prior range, and the next rise begins from a cleared state. When it does not, activation carries forward.

This restorative capacity develops through repeated experiences of recovery — return to rhythm, coherence, and a level of activation the system can sustain.

The conditions that support recovery are not only internal. They include environmental, task, and relational factors that shape how the system can return toward baseline. In many autistic profiles, these conditions are more specific. Environmental pacing, sensory load, and predictability may support or interfere with recovery. Co-regulation can support return toward baseline when the form it takes matches the system's processing conditions. When it does not, co-regulatory signals may increase concurrent demand rather than reduce it. Within this framework, prosodic, rhythmic, or relational input that requires precise prediction conditions to process may add load rather than clearing it, consistent with accounts of atypical precision-weighting in autistic prediction (Lawson et al., 2014). This specific effect of co-regulatory input has not been directly tested.

8.4 — Overload: When the Slope Exceeds Available Margin

Overload is an acute threshold event — the point at which activation exceeds available margin in the present moment.

When activation exceeds available margin, protection begins to organize. In autistic bodies, this may appear as meltdown or shutdown — outward discharge or inward withdrawal (Mazefsky et al., 2013).

The threshold crossing is physiological before it is behavioral.

Regulatory Field in Transition — Overload

A shift begins.

Muscle tone increases. Breath shortens. Sensory edges sharpen. Slope is rising. Concurrent demand is increasing.
The room has not changed. What registers has changed. Signals that had remained peripheral move forward. Integration requires greater effort.
Activation rises incrementally. The system works to preserve continuity.
Concurrent demand exceeds what current organization can stabilize.
Language becomes less immediate. Processing slows as resources are reallocated.
The regulatory margin thins.
Internal rhythm destabilizes. Degrees of freedom decrease.
A threshold is crossed internally before it becomes visible externally.
Protection is not chosen. It emerges.

Overload marks the point at which sustaining openness exceeds available margin. The system shifts toward protection to preserve viability — the cost of continuing at the current level of functional engagement has exceeded what remains available. Discharge may reduce physiological intensity through outward expression. Shutdown may reduce immediate concurrent demand by withdrawing from engagement. Within an RNS reading, both may be understood as different protective responses associated with a threshold dynamic in which current regulatory demands exceed what the system can sustain.

When recovery between threshold crossings is incomplete, the consequences of these events do not remain local to the moment — they accumulate across cycles. That extension is what the next section traces.

8.5 — Regulation and Energetic Availability

The same person may regulate fully in one moment and lose functional access in another without any change in capacity, intention, or learning. What differs between those moments is the physiological state from which the system is operating — the regulatory margin and the energetic availability present at that time.

What a person can produce varies — and that variability is legible. Available margin fluctuates with how completely the system has recovered and how well the environment's demands fit the system's parameters. When recovery has been insufficient or demand has been poorly matched, the system operates from reduced margin, and what was possible is no longer available.

The same task that was within easy reach last week may require genuine effort this week because the system is operating from an elevated baseline shaped by insufficient recovery across the intervening days — a pattern consistent with the within-individual performance variability documented in autistic populations (Karalunas et al., 2014).

Increasing demand on a system with reduced energetic availability does not necessarily improve output. It may instead raise the physiological cost of each attempt, move the system faster toward overload, and make recovery less likely — because the conditions that would restore margin are consumed in the attempt.

Not all accumulation is environmental. Load may originate within the system itself — disrupted sleep, unresolved pain, unmet physiological need — functioning as concurrent demand before any external condition is encountered. A body managing such internal load arrives at every environmental demand already carrying what has not been cleared.

Reading this accurately begins from the state the system is actually in and what is generating it — not from the output the system is producing. The system does not recover through instruction. It recovers when conditions allow recovery to proceed.

Accumulation follows two distinguishable pathways. The first is visible: repeated threshold crossings that leave residual activation across cycles. The second is less visible: sustained adaptive organization under conditions of continued demand, where activation is held within functional limits but does not fully resolve — often mistaken for stable functioning.

What is altered across time is not only the level of load, but the system's capacity to return — the regulatory rhythm that would otherwise allow each cycle to close.

8.6 — Exhaustion: The Floor That Has Been Rising

Exhaustion reflects the prolonged expression of this second pathway—what the system becomes when sustained activation has continued without full return across time.

The system is operating from a state that prior recovery has not restored. This is not a more severe version of overload. It is a different condition — one in which the floor itself has become the new baseline.

Paired Scenario — Reduced Margin, Present State

It is Thursday afternoon. The floor has already risen. This is not a prediction about what will happen — it is a description of where the system is now.

For the first child, recovery across prior cycles has been sufficient. Baseline has risen and cleared incrementally. Margin remains available.

For the second child, the floor has not cleared. The system is operating from a baseline that has not returned. This is not visible in posture, output, or engagement. It is present in what any new demand will cost.

The same perturbation arrives.

The child whose floor has cleared: slope rises within available margin.

The child whose floor has been rising: slope rises into a system whose margin has already been consumed.

The arc does not produce the elevation. It encounters it.

The event did not produce the response. It encountered a system already organized at reduced margin.

"It's not that something happened. It's that everything kept happening and nothing was ever enough to recover from it." One autistic participant in Raymaker et al.'s (2020) study described it plainly. Another described the loss of previously available capacity as one of its most alarming features: "I couldn't do things I used to be able to do. Basic things. And I didn't know if they would come back."

Within the framework developed here, these accounts can be read not simply as a threshold crossing, but as a configuration the system has arrived at — one in which ordinary demand may exceed what available margin can sustain. This reading does not resolve whether the underlying capacities themselves have changed; it describes the regulatory conditions under which those capacities are currently available. Figure 20 maps the emergence of chronic load.

Figure 20 — How Chronic Load Emerges Across Repeated Cycles

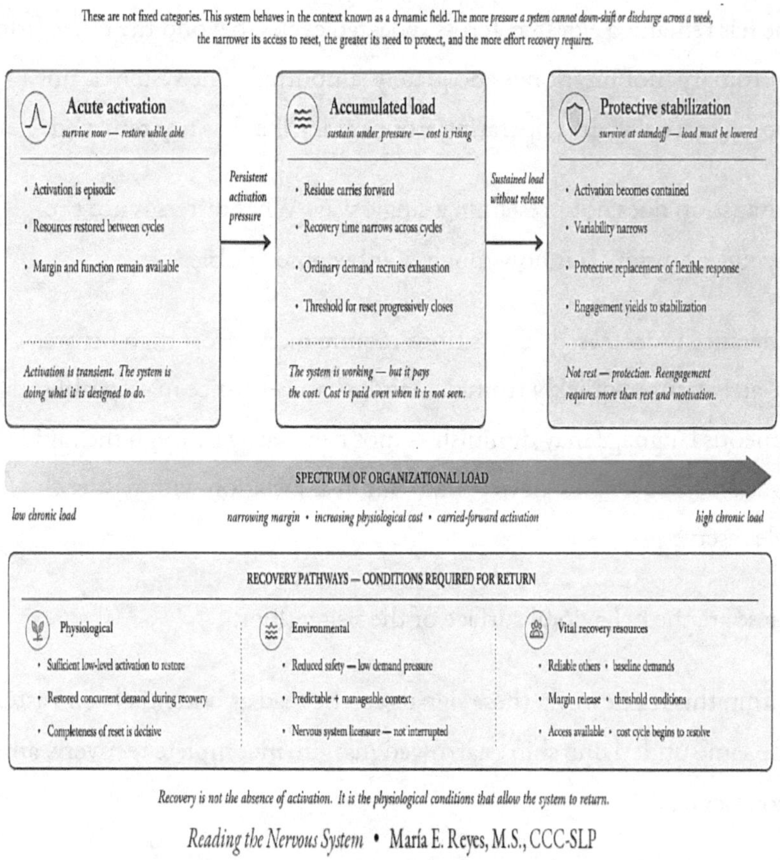

This figure shows the baseline the system is now operating from — not a single activation event, but the floor that has formed across prior cycles.

This distinction aligns with the empirical literature on autistic burnout, which describes a state of sustained depletion distinct from acute overload: reduced capacity across domains, withdrawal from participation, and recovery timescales measured in weeks or months rather than hours (Raymaker et al., 2020; Higgins et al., 2021; Mantzalas et al., 2022).

Exhaustion is not a more severe version of overload. Within this framework, exhaustion can be read as a condition in which the regulatory baseline has remained elevated across repeated cycles, beyond the range from which ordinary mobilization is adequately supported. The system is not responding to a spike. It is operating from a floor that has not returned.

Exhaustion does not present in a single way. What differs is not the underlying condition, but how much of it becomes visible.

Tolerance is decreased. Transitions require more effort than before. Fatigue arrives more quickly from demands that were once manageable. Spontaneous language may diminish — not necessarily through the rapid rise in activation characteristic of overload, but in association with reduced available reserve.

These are the behavioral surface of the rising floor.

Within this framework, these signals can be read as potentially consistent with the same underlying shift: narrowed margin, incomplete recovery, and reduced reserve.

The asymmetry this creates is clinically significant. Within this framework, recovery from an elevated baseline may be slower and may require more supportive conditions than recovery from a more restored baseline. The system becomes harder to restore precisely as the need for recovery increases.

Closing — When the Body Needs to Restore

When energetic resources are adequate, the system can move through activation and return toward baseline. When they are not, it protects what remains.

Within this framework, overload, shutdown, and exhaustion can be read as coherent expressions of a system reorganizing toward protection when the cost of remaining open exceeds what is available.

The signals that precede these states are often present before they become behavioral: rhythm shifting, margin thinning, conservation beginning. Reading these signals allows adjustment at the level of conditions rather than behavior — adjusting demand, restoring predictability, and protecting recovery time before threshold is crossed.

Regulatory variability does not define autism. It reflects how nervous-system organization is shaped by the fit between individual processing patterns and the conditions the environment requires.

Regulation is not only a present-moment phenomenon. When activation does not fully resolve, a specific residue remains. It is not narrative memory. It is a slightly elevated baseline from which the next experience begins.

That residue becomes the condition through which the next experience is felt.

Chapter 9 — Emotion and Bodily Memory in Autism

Opening — What the Body Carries Forward

What Chapter 8 traced in activation and residue, this chapter traces in emotion: what the body carries forward, and what that carrying forward does to present organization.

Emotion is approached here not as psychological content or expressive skill, but as a primary bodily process. It follows the same arc of activation, integration, and recovery traced in the preceding chapters, and leaves residue when that arc does not reach completion. What remains is not the event itself, but the configuration through which the experience was lived: how much activation it required, whether continuity was preserved, whether recovery occurred, and whether the body moved through that arc in accompaniment or alone.

Within this framework, emotion is read not as originating in an isolated event, but within the continuity of signals that have been completed, interrupted, or sustained without resolution.

The account that follows represents the framework's most inference-dependent extension. It is also its most phenomenologically vivid. The two are related. The sequences that follow may feel concrete and lived; their epistemic status is more tentative. The inferences are theoretically grounded and consistent with outcomes documented in the empirical literature, but they are not demonstrated mechanisms. The reader is asked to hold that distinction.

The term *bodily memory,* as used in this and the following chapter, is a descriptive construct rather than a reference to an established memory system or a claim about a discrete neural mechanism. It refers to the retention of physiological organization across time — how prior patterns of activation, completion, interruption, and accompaniment continue to organize present expectation through changes in baseline, threshold proximity, and anticipatory organization. The concept draws on broader work in embodied cognition and somatic memory (Damasio, 2010; Fuchs, 2012), while applying those ideas within the regulatory framework developed here. The specific formulation advanced here is theoretically grounded but not yet independently verified in autistic populations.

9.1 — How Environment Shapes Internal Reading

The developing regulatory system does not build its interoceptive organization in isolation. Internal signals are not registered in isolation either. They are registered alongside the external conditions that repeatedly accompany them — sound, movement, proximity, temperature, relational stability — and the nervous system builds its anticipatory patterns from those repeated pairings.

What is carried forward is not the event itself, but the physiological organization through which it was lived. Interoceptive thresholds, pacing of integration, and baseline reflect these accumulated pairings. The architecture remains autistic. What the developmental environment shapes are the anticipatory patterns that form within it.

These patterns are not static. As each cycle resolves or remains incomplete, the organization updates — shifting the baseline from which subsequent encounters begin.

9.2 — Emotion as a Primary Bodily Process

Within this framework, emotion at the level of physiology is approached as modulation of activation, rhythm, and availability — a shift in the regulatory field that can be read through the same parameters used to describe other activation events.

At the level of experience, emotion reflects what it is like to move through that arc. This reframes the question from "why is this person emotional?" to "what conditions are present for the arc to reach completion?" The patterns described here follow physiological conditions — repeated activation, incomplete recovery, and the presence or absence of a regulating field — rather than interpretation or psychological content.

Within the framework, repeated activation arcs that reach sufficient completion are expected to contribute to recalibration, while repeated incomplete arcs may contribute to accumulation.

9.3 — Emotional Arcs, Residue, and Carried Load

Emotional activation does not necessarily resolve within the interval available before the next demand arrives. When descent extends beyond that interval, activation persists as residue into what follows.

This carried activation is not inert. It continues to shape perception, response, and the conditions under which subsequent activation unfolds. Each new arc begins in the presence of what has not yet completed, altering both its starting point and its trajectory.

What began as a transient shift becomes the condition from which subsequent activation proceeds.

In this context, mismatch is not only what increases load. It is the condition under which emotional processes fail to integrate, and through which what remains begins to organize future anticipation.

9.4 — Accumulation and the Emergence of Sensitivity

When emotional processes repeatedly fail to integrate, cost accumulates through the same dynamic described in Chapter 8 — each incomplete recovery leaving a small increment within baseline, narrowing regulatory margin in advance of each subsequent demand. This accumulation tends to occur more rapidly in systems carrying higher concurrent demand, longer recovery times, and more specific conditions for emotional completion.

What that accumulation produces, across cycles, is a shift in where the system is operating relative to threshold.

Sensory responsiveness is not necessarily expressed as a fixed property across conditions. Its moment-to-moment expression may also vary with regulatory state.

Within RNS, one possible contributor to that variation is the system's current position relative to threshold: as baseline rises and available margin narrows, inputs that were previously manageable may be encountered under conditions of reduced regulatory range. This is a framework-level interpretation of state-dependent variability, not a claim that regulatory margin determines sensory thresholds.

Within this framework, heightened sensory responsiveness may in some contexts be read partly in relation to threshold proximity and current regulatory conditions, without reducing hypersensitivity to those conditions or excluding relatively stable sensory-processing differences.

The same regulatory logic offers one possible account of state-dependent reductions in responsiveness. Under prolonged activation or sustained protective organization, responsiveness to some inputs may change. In some individuals and contexts, greater stimulus intensity may be required before a response becomes observable. This should not be read as an explanation of sensory hyposensitivity itself, but as a framework-level account of how current regulatory conditions may influence the expression of responsiveness.

Within an RNS reading, reduced responsiveness can therefore be examined as a changing relationship among incoming signal, current state, and available processing conditions—not assumed to represent an absence of sensitivity.

Hyper- and hyporesponsive states may, in some individuals and contexts, be influenced by regulatory state without implying that a shared regulatory mechanism explains sensory hyper- or hyposensitivity. The same person may show different patterns across conditions as baseline, concurrent demand, recovery, and available margin change. State-dependent variation may therefore be layered onto more stable sensory characteristics.

Sensory processing is one domain in which state-dependent shifts may become especially observable. Sound tolerance, light tolerance, and tactile responsiveness can vary with context and current state; within RNS, available margin is one condition considered when interpreting that variability. This does not reduce sensory-processing differences to regulatory margin or treat perceptual thresholds as direct measures of it. A sound that was manageable earlier in the day may be experienced differently after sustained demand: the stimulus is similar, but the conditions in which it is being processed are not.

When sensitivity is treated only as a fixed trait, attention to possible state-dependent variation can be lost. RNS therefore treats regulatory state as one layer of interpretation alongside sensory-processing characteristics, context, history, and first-person report when available.

What is carried forward may include not only activation but learned and state-dependent conditions that influence how subsequent signals are encountered. Section 9.5 considers this possibility in relation to bodily memory.

9.5 — Emotion, Stress, and Relational Bodily Memory

Current input may engage previously established patterns of activation, anticipation, and recovery — not through recall of the prior event, but through alignment with its configuration. The system responds to similarity of state rather than similarity of narrative: what is activated is not a remembered event, but a recurrent physiological arrangement — the same activation level, the same margin, the same absence of accompaniment. This account is an inference from the framework. It is consistent with documented outcomes but not yet directly tested as a mechanism.

When the configuration of a present moment aligns with a prior pattern of incomplete activation or unaccompanied arc, the system may respond to that alignment before conscious interpretation has organized around it. In this way, prior organization shapes how internal signals are registered in the present — what is available, what is suppressed, and how much demand is already being carried before anything new arrives.

What shifts is the starting condition: the baseline from which the system proceeds, the proximity to threshold it carries into relational encounters, and the concurrent demand already active before any new signal arrives.

Within this reading, the response need not be understood as originating entirely in the present moment. Prior organization may contribute to the conditions from which the present response emerges.

9.6 — When Emotion Was Not Accompanied

Most activation arcs resolve without external input. The body recalibrates.

Difficulty emerges when activation repeatedly unfolds without sufficient conditions for completion. Under these conditions, the system begins to organize around the likelihood that activation will not resolve fully.

This is not avoidance of emotion. It is allocation of resources based on prior experience of cost and incomplete recovery.

As this pattern stabilizes, the system does not only carry activation forward — it begins to reorganize in ways that reduce exposure to similar conditions. This transition, from accumulation to reorganization, marks the emergence of protective organization.

The shift is qualitative, not merely quantitative. The system is no longer only accumulating load from incomplete recovery. It has begun to organize in advance of that load, shaping response before activation fully unfolds.

At the level of the four parameters, baseline has elevated to the point that threshold proximity becomes the system's dominant organizing condition. Figure 21 shows this consolidation across cycles.

Figure 21 — Emotional Completion, Accumulation, and the Emergence of Protective Organization

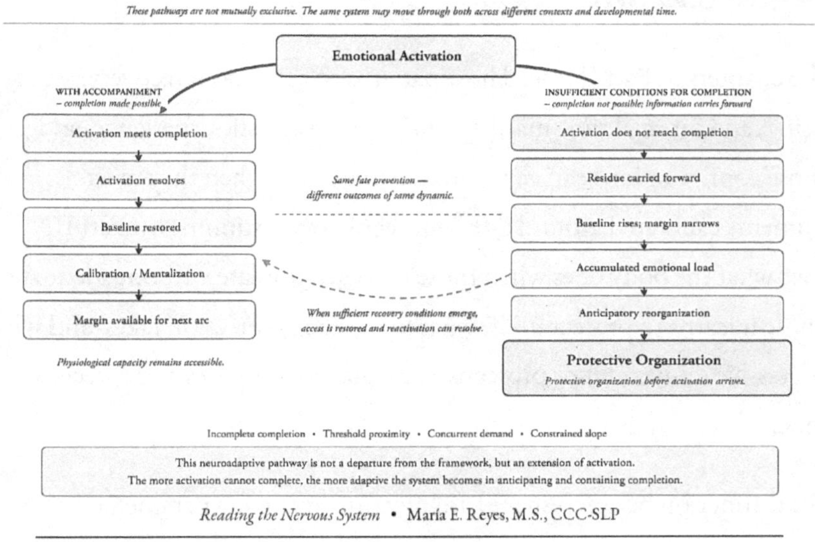

Within this framework, emotional activation is treated as a physiological process that unfolds across time and whose completion may depend on sufficient regulatory conditions. When these conditions are consistently unavailable, adjustment follows cost.

What persists is not the event, but the conditions under which experience has organized — and from which anticipation takes shape. This is not memory as recollection, but memory as organization.

Within the proposed account, what begins as a response to activation may become an organization that anticipates it.

Part III traces what the body does next: how anticipation becomes architecture, and what that architecture requires and forecloses across time.

PART III — Safety, Protection, and Bodily Reorganization

The chapters in Part II traced how patterns of activation, recovery, sensory processing, and internal time may be configured in autistic experience, and how those configurations can generate recurring costs when they meet environments calibrated around different regulatory assumptions. Part III examines what the body does when those costs accumulate without adequate relief: how it learns to protect itself, what forms that protection takes, and what the progressive refinement of protective organization requires and forecloses over time.

The distinction between shared architecture, recurring parameter tendencies, and state-dependent configurations remains operative throughout what follows. It matters with particular force here. The protective progressions traced across these chapters are state-level and history-level consequences: patterns that may emerge over time in an autistic nervous system under sustained conditions of mismatch. They are not defining features of autism itself. Reading them accurately requires holding the distinction between what the system is and what sustained mismatch has required of it.

What the system organizes toward is not the anticipation of events, but the anticipation of the signals those events have come to generate. Protection forms around what the body has learned will arrive — not as a discrete occurrence, but as a pattern of activation it must be prepared to sustain.

The account that follows does not describe all autistic experience or imply a uniform developmental trajectory. It describes a recognizable progression — physiologically coherent and documented across research and first-person accounts — that a nervous system may follow when repeated activation without adequate recovery shapes its anticipatory organization. Each form of protection is rational given what the system has learned, and each carries costs that become visible only long after the adaptation has established itself.

Chapter 10 — Safety and Protection: A Reading from the Body

"I mask so well sometimes that people don't even know I'm autistic until I give them the big reveal. But at the end of the day, it gets really stressful. Even something like doing my homework becomes very tiring." — Ethan Lisi, 2020

"It's exhausting pretending to be someone you're not for the majority of every single day. You get so used to performing that you forget there was ever anything to perform." — Sarah, late-diagnosed autistic woman (Bargiela, Steward, & Mandy, 2016)

Opening — When the Body Prioritizes Continuity

Every interaction required full attention — not to what was being said, but to whether the response was correct, the face was doing the right thing, and the right amount of space was being taken. Listening and performing could not occur at the same time. The performance was always running. It was not recognized as performance until much later.

By the end of a workday, there was nothing left. Not tired — emptied. The commute home was the only time the performance stopped, and with it came the question: what was actually there when no one was watching? Sometimes the answer didn't come.

These are not isolated descriptions. Within this framework, they can be read as converging descriptions of a similar regulatory problem from within: a system that had learned to sustain outward organization while containing internal signals — not as a choice, but as the most viable option when open expression had come to cost too much.

Chapter 9 closed with a precise movement: protection, which began as a response to activation, had learned to anticipate. When the system can neither withdraw nor express without cost, it sustains outward organization while containing internal activation. This chapter begins from that point.

It traces what happens when anticipation becomes the system's organizing mode — not as an episodic event but as structural baseline: how the alertness that was once a response to demand becomes the resting level from which all demand is met; what it costs to sustain that alertness without adequate recovery; and what form protection takes when open defense is no longer viable and withdrawal is not possible.

After sustaining more than it can recover, the body does not stop. It shifts.

10.1 — Safety as a Bodily Condition

Safety is not only an idea. Within RNS, sufficient regulatory margin is treated as one physiological condition that can support openness to experience without the cost of that openness exceeding what recovery can restore. At the level of experience, safety may be felt when the body does not need to organize immediately toward self-protection: when signals can arrive, be registered, and settle while functional continuity remains available.

Protection, as used throughout this chapter, does not refer to a psychological strategy or behavioral choice, but to a physiological reorganization that preserves functional continuity under constraint. It is not chosen. It emerges. Rather than increasing activation, protection reorganizes the system in advance of it — constraining variability, prioritizing predictability, and allocating resources to maintain continuity when recovery cannot be assumed.

Clinically, this may appear as the difference between a nervous system that can be reached — one that allows approach, accepts support, and remains present through difficulty — and one whose organization is consistent with an expectation that remaining open carries greater regulatory cost.

Within this framework, aspects of safety appraisal may occur sub-symbolically, before conscious interpretation. The system registers tone, rhythm, intensity, predictability, and congruence — what neuroception names (Porges, 2011, used here for descriptive utility rather than as a commitment to polyvagal theory's more contested mechanistic proposals, as noted in §4.8) — and reorganizes according to whether what is arriving resembles conditions that once permitted recovery or conditions that did not. "Nothing is wrong" does not alter the response if bodily signals remain inconsistent with safety. Verbal interpretation is understood as only one influence on the process; physiological state also shapes how incoming conditions are registered.

In some autistic profiles, this process may appear more immediate or less attenuated — with less mediation between signal detection and organizational consequence. Subtle environmental variations may be detected and responded to before conscious interpretation can mediate the response. It reflects a bodily appraisal refined across time: a more detailed reading of signal in an architecture that relies more heavily on internal precision and less on contextual interpolation.

Bodily availability, in this frame, is present when the nervous system can sustain shifting activation while preserving functional coherence, recover after effort before the next demand arrives, tolerate uncertainty without immediate defensive shift, and move through intensity without losing internal continuity. As margin narrows — as concurrent demand rises, recovery becomes incomplete, and baseline elevates — these capacities become progressively constrained as the system recalibrates within the available regulatory range. With adequate internal margin, other capacities expand: curiosity, flexibility, exploration, and relational availability. These emerge as resources that basic protection no longer requires. What appears psychologically as openness or engagement may therefore also reflect changes in bodily availability.

Within this framework, a more durable sense of safety may develop as the nervous system accumulates repeated bodily evidence that remaining open under particular conditions does not reproduce the regulatory cost associated with prior protective organization. Verbal reassurance alone may not be sufficient. The framework therefore emphasizes repeated experience in which remaining present is compatible with available margin and recovery is possible afterward.

10.2 — From Momentary Response to Protective Organization

The progression described below is an RNS formulation of how repeated regulatory cost may become anticipatory organization. Learning, prediction, autonomic adaptation, and experience-dependent plasticity provide established foundations for such a possibility; the specific sequence proposed here has not been demonstrated as a single physiological pathway.

Protection functions best when it is temporary. The body detects risk, adjusts, and reopens as conditions shift. Difficulty begins when the margin for reopening is no longer available. What begins as a response to specific conditions may, through repetition and incomplete recovery, organize into an anticipatory posture from which all subsequent conditions are met.

Within the proposed progression, repeated experiences of unaccompanied activation may shape not only how the system responds but what it anticipates. This anticipation is not framed as a conscious expectation, but as a possible physiological calibration shaped by repeated conditions across time. RNS proposes that the system may become increasingly organized around signal configurations that prior experience has associated with regulatory cost. Chapter 9 introduced bodily memory as a descriptive construct for the carried influence of prior activation, completion, interruption, and accompaniment. When similar configurations recur, anticipatory organization may begin before the full demand arrives.

Regulatory Field in Transition — From Response to Baseline

Activation rises. Muscle tone increases. Breath shortens. Attention narrows. The event passes. Recovery begins but does not complete. Intensity lowers partially. Variability does not fully return.

Another demand arrives before margin has restored. Activation rises again, now from a slightly elevated resting level. Return slows. Restoration requires greater effort.

The cycle repeats.
The interval between activation and reopening shortens. Anticipation begins to precede intensity. Alertness stabilizes. What was once transient becomes organizing.

Margin remains narrowed — not because danger is constant, but because reopening has repeatedly carried unsustainable cost. Protection settles into baseline.

Within this formulation, reopening may require conditions robust enough not only to signal safety but to support energetic viability. A cue of safety, by itself, may not be sufficient. Repeated embodied experience in which remaining open carries lower regulatory cost is proposed as one condition through which anticipatory organization may begin to change.

10.3 — The Cost of Sustaining Alertness

Fatigue, rigidity, or withdrawal are often interpreted as lack of motivation, resistance, or limited capacity. Read from the body, they may instead be consistent with a system that has been working continuously to preserve access. Within this framework, sustained protection carries *cumulative physiological cost.*

Regulatory Field in Transition — Sustained Alertness

There is no threat present.

The room is ordinary. The people in it are known. Nothing has changed.
And yet.

Shoulders remain slightly raised. The back of the throat is slightly tighter than it needs to be. Breath moves at a depth just above full. Monitoring continues.

This is not a response to anything happening now.

It is the cost of having learned that readiness is safer than the alternative.
The cost is not visible. The cost is continuous. The cost is being paid right now, before the day has asked for anything.

When alertness stabilizes as baseline, activation may remain elevated even without visible external demand — a form of rising floor related here to the continuing resource expenditure associated with sustained vigilance.

Over time, cost becomes visible in function. Tolerance decreases. Recovery slows. Engagement narrows. Activities that were once manageable begin to feel heavy — the resources required for capacity are already committed to vigilance. Within this framework, collapse can be read as the delayed expression of accumulated regulatory expenditure that ordinary intervals of recovery may not have fully compensated.

Demands that are added at this stage — greater flexibility, increased adaptation, expanded performance — further reduce remaining margin. What appears as noncompliance may reflect physiological depletion.

What appears as rigidity may reflect a system preserving the predictability that keeps adaptation cost from exceeding what is available. A body sustaining alertness is a body working without rest, producing what it can from a base that has been emptying over time.

10.4 — When Lowering the Guard Is No Longer Possible

This section describes a further development: the point at which the system loses flexible movement between protected and open states.

This occurs when the threshold for protective response has been organized by bodily memory in a way that shapes how many environmental variations are encountered, including whether they are associated with increased caution. The organism sustains baseline alertness even in contexts that appear safe. Anticipation shifts from an occasional resource to an organizing principle of daily functioning. The system no longer easily distinguishes between contexts that were once demanding and those that merely resemble them — not because the capacity to discriminate has disappeared, but because prediction is now shaped by traces built from repeated experiences of activation that completed without accompaniment.

Regulatory Field in Transition — When the Guard Cannot Lower

The room is familiar. The people in it are known. The day has asked for nothing unusual.

The signal arrives that conditions are safe. The body receives it. The body does not lower.

Not refusal. Not disbelief. Within this reading, prior learning may constrain how strongly a current safety signal reorganizes the system. Safety, registered. Guard, maintained. The two may remain partly uncoupled.

What arrived as reassurance was processed as information. It did not become a condition.

Externally, this state may be barely visible. A child may appear attentive, cooperative, or calm while internally sustaining elevated physiological cost. Behavioral calm may coexist with reduced internal access. When this pattern persists, flexibility between activation and recovery diminishes. Variability narrows. Recovery becomes fragile.

At this point, a consolidated protective configuration shaped by bodily memory can be described — not as diagnosis, but as functional formulation: a way of preserving continuity after repeated experiences in which environmental, sensory, or relational load exceeded integrative capacity. This is not regulatory failure. It is regulation that has succeeded, for a long time, under conditions that required it to take this form.

Relational Consequences of Protective Organization

Protection does not only reorganize internal states — it alters how other nervous systems are perceived and anticipated. Within this framework, a system organized around sustained protection can be understood as learning to predict that authentic expression carries relational cost under conditions similar to those in which the pattern developed. This prediction shapes not only whether the system remains open, but how it interprets relational signals from others.

Within the proposed model, established protective organization may bias the processing of incoming relational information toward conditions associated with prior regulatory cost, even when current relationships offer different possibilities.

The figure that follows is not read as a checklist of stages, but as a progression of protective logic — each form emerging from the conditions the previous one could no longer sustain. Read it as a directional trajectory: what the system learned to do, as the cost of remaining open accumulated. As §10.5 clarifies, these are not exclusive or strictly sequential stages — a system may employ more than one form at once, or shift among them out of this order. Figure 22 traces this progressive consolidation.

Figure 22 — Progressive Consolidation of Protective Organization

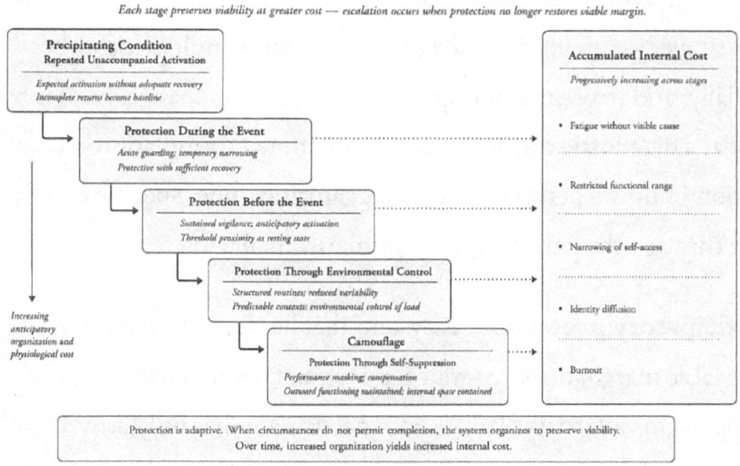

This figure is not a model to be applied. It is a simplified reading of a process that is already occurring continuously.

Reading the Nervous System • María E. Reyes, M.S., CCC-SLP

The forms described below are not separate categories but progressively consolidated expressions of the same protective logic, each emerging when earlier strategies no longer sufficiently restore viable margin.

10.5 — Forms of Protection: Conceptual Architecture

Protective organization does not take a single form. Once lowering guard becomes physiologically costly, protection may organize through several distinct strategies depending on what the environment permits.

What follows names three distinct forms that protective organization may take. Within this framework, they can be read as different responses to a shared regulatory problem — how to remain functional when staying open carries greater cost than the system can readily sustain — while remaining behaviorally and experientially distinct. These forms do not appear suddenly. Each consolidates gradually through repeated conditions across time, as earlier regulatory responses prove insufficient to restore viable margin.

The configurations below represent distinct physiological arrangements managing different parameters under sustained load. They are not exclusive types. A single system may employ more than one, simultaneously or sequentially, and may shift among them as conditions change. Each is named here not as a diagnostic category but as a functional formulation: a compact description of how a nervous system is organizing under specific conditions, readable through the parameters it is primarily managing.

Anticipatory protection. A system that has learned that activation often exceeds viable margin does not wait for overload before protecting itself. Guarding begins internally, before external demand has fully arrived. Subtle shifts in tone, context, or sensation are enough to initiate adjustment. In the terms of Part I, regulatory margin has narrowed and threshold proximity has increased. Signals that would pass in a wide-margin system now recruit activation. Internal time moves ahead of experience. Response is organized around what is predicted rather than what is fully present. Protection may activate earlier as threshold proximity increases following prior activation that has required unsustainable recovery.

From the outside: anticipatory anxiety, rigidity, or rapid defensive adjustment. From the inside: activation recruited early, before arcs the current margin cannot contain. This form of protection is organized around prediction. It reduces arousal before accumulation.

Protection through reduced exposure. A system that has learned that activation often exceeds viable margin may adapt not only internally but environmentally. Rather than relying primarily on early prediction, it modifies its surroundings. Exposure is reduced through structured routines, limited variability, shaped transitions, and predictable contexts. Environmental load is reduced to the point that margin remains viable and the floor does not continue rising, because processing cost has been brought low enough for recovery to occur between demands.

Predictability becomes a way of preserving continuity across activation shifts. Variability is constrained because it has repeatedly generated activation arcs that available margin cannot contain. Within this reading, reliance on routine need not be understood as mere preference; predictable environments may carry lower regulatory cost than variable ones.

Within an RNS reading, routine may function as one way of shaping environmental conditions so that demand remains more compatible with available regulatory margin.

From the outside: strong preference for predictability, structural reliance on routine, or resistance to variation. From the inside: the system maintaining viability by reducing the unpredictable. This form of protection centers on environmental control. Reducing variability preserves regulatory range.

These two forms of protection are, at least in part, externally legible. Anticipatory protection may appear as rigidity, rapid adjustment, or what is often described as anxiety. Environmental control may appear as strong preference, insistence, or resistance, and is often misidentified as personality-based rigidity rather than recognized as regulatory architecture.

A third form of protection is less visible. It emerges when neither internal guarding nor environmental control is sufficient — when the system must remain present and cannot reduce the conditions it is managing. That form is the subject of the section that follows.

10.6 — Camouflage: When Protection Becomes Invisible

The account that follows assembles a proposed pathway from converging inferences. Each step is internally consistent with the framework and aligns with the empirical outcomes the literature documents. What it does not provide — and does not claim to provide — is a mechanistic demonstration of the pathway itself. Camouflage, and the identity consequences it produces, are empirically established phenomena. The physiological route proposed here is one account of how they arise: a theoretically grounded explanation of established outcomes, not a closed causal demonstration.

The account draws on published autistic accounts as primary evidence of lived organization, while recognizing that it has not been co-developed with autistic collaborators — a limitation that bears particular weight here, where the framework is most directly proposing a developmental pathway through autistic experience. The gap between reading from autistic accounts and building with autistic collaborators is real, and future development of this framework should address it directly.

Camouflage, as described here, is not the adoption of a false self, but the maintenance of outward organization while internal signals are contained under conditions where open expression carries physiological cost.

If the organism cannot lower its guard yet cannot sustain open defense, another form of safeguarding may emerge — quieter, more elaborate, and often invisible from the outside.

This is camouflage.

This form of protection is a functional refinement of protective organization. It arises when overt defense is no longer viable and withdrawal is not possible. The organism must remain present without exposing itself directly or collapsing.

The Preconditions

Chapter 9 traced the transition from which camouflage emerges.

When two physiological learnings converge — that full emotional activation in relational contexts is costly, and that expression through conventional relational channels tends to increase concurrent demand rather than reduce it — the system arrives at a precise economy: sustaining outward organization while containing internal signal carries lower overall cost than either open expression or withdrawal. Within this framework, the economy is physiological, not strategic — an accumulation of what each available option costs.

The cost-differential described in §4.8 and §8.3 applies directly here: for systems with different social-processing architecture, the relational signals that constitute co-regulation may carry higher processing cost rather than reducing load.

Through sufficient repetition of this pattern, the system appears to register something precise: being seen as one feels does not reliably reduce risk. Conventional relational channels do not consistently reduce load.

They often increase it, even when the intention is support.

Two conditions make camouflage the solution the system arrives at rather than withdrawal or open defense. First, overt defense is no longer viable: the responses it evokes — from others, institutions, or relationships on which the system depends — have been learned to carry costs greater than containment.

Second, withdrawal is not possible: relational presence is required by the environments the system must inhabit. When both conditions hold, and when conventional expression has been learned to increase rather than reduce load, sustained outward organization becomes not a choice but the most viable physiological solution available.

This is the condition from which camouflage emerges: not from a decision to conceal, but from the physiological pattern that spontaneous expression through conventional relational channels tends to increase concurrent demand rather than reduce it.

The Mechanism

Instead of withdrawing, expression is modulated. Gestures are adjusted, spontaneous movement inhibited, emotional display regulated, and internal signals contained. External expression aligns with expectation not because the expectation is accurate, but because misalignment has previously produced responses that cost more than alignment does.

Hull et al. (2017, 2020) identified three broad components of camouflaging in autistic adults: assimilating, masking, and compensating.

Within the framework developed here, these can be read as behaviorally and experientially distinct expressions of a shared regulatory process: the continuous management of outward presentation, which may reduce the activation cost of social environments, at the price of an internal labor the surface does not reveal.

Camouflage carries measurable diagnostic and clinical consequences. Higher camouflaging scores are associated with lower autistic trait scores on standard assessments, greater rates of missed or delayed diagnosis, and worse mental health outcomes — including anxiety and depression — with especially marked effects in autistic women and gender-minority individuals (Hull et al., 2020). In many such cases, late diagnosis reflected not the absence of autistic characteristics but their successful concealment from external observation, including clinical observation.

In many contexts, camouflage allows presence without excessive exposure. It permits participation in social, institutional, or relational environments while maintaining outward functioning aligned with expectation, without activating defenses that would increase energetic cost or trigger responses the system has learned to avoid.

This form of safeguarding often goes unnoticed. From the outside, a person may appear regulated, adapted, or functional. Internally, effort remains high. Continuous monitoring is required: anticipating implicit demands, correcting spontaneous responses, and sustaining an outward organization that does not match internal state.

Camouflage does not reduce arousal. It redirects it inward. What might once have been expressed through discharge, rigidity, or withdrawal becomes contained, drawing on available resources to sustain control and self-correction. Experience does not disappear. Emotion remains present, though with fewer channels for expression. Closure is deferred.

What once moved outward is carried inward.

This form of protection emerges because it works. It allows continued presence, reduces negative consequence, and preserves connection when other protective forms are no longer viable. Read from lived physiology, it is a sophisticated response of a nervous system that has organized itself around remaining present without being seen.

Its sophistication is not a measure of its cost. It is precisely because camouflage works — because it preserves outward functioning and reduces visible disruption — that its cost remains hidden and the conditions producing it often remain unchanged.

The Labor and Its Cost

This organization functions not only as relational adaptation but as sustained internal labor. What appears externally as composure often requires continuous micro-adjustments in posture, facial expression, vocal modulation, timing, and inhibition of spontaneous movement. Each adjustment carries energetic cost. Maintained across hours, days, or developmental years, the cost accumulates.

As the participant quoted at this chapter's opening described it (Bargiela et al., 2016): every interaction required monitoring whether the face was doing the right thing, whether the right amount of space was being taken. Listening and performing simultaneously was not possible. The performance was continuous.

The second account quoted at the chapter's opening (Bargiela et al., 2016) names the cost in equally revealing terms: nothing left at the end of the workday, not tired but emptied, the commute home the only interval without performance. The phrase "what was actually there when no one was watching" names a consequence that extends beyond fatigue: not only depletion of resources, but progressive loss of reliable access to one's own states — what §9.5 identified as the narrowing of self-reading produced by sustained adaptive management.

Unlike overt protection — withdrawal, rigidity, visible activation — camouflage absorbs activation inward. Arousal does not discharge through visible response; it is contained through effortful control. Restoration may occur only partially. Because outward functioning is preserved, environmental demand often remains unchanged. The system receives no signal to reduce load.

Over time, this pattern reshapes regulatory margin. Energy that might otherwise support exploration, creativity, relational exchange, or recovery is redirected toward monitoring and self-correction. As this redirection persists, internal access narrows. Fatigue increases without visible breakdown. What remains active within the system limits what is available for recovery. The accompanying person attending only behavioral output would see a person managing well. The system itself is running in progressive deficit.

The Identity Dimension

The narrowing produced by camouflage involves, on this account, more than fatigue or reduced relational availability. It extends to self-knowledge — to the system's ability to read its own states as distinct from the managed presentation it has been sustaining.

Section 9.5 traced how bodily memory shapes self-reading: how accumulated patterns of containment and self-correction become part of how internal signals are organized, such that what the system registers in the present is shaped by the history of how similar signals had to be handled. When that history includes extended periods of self-monitoring and expression management, the distinction between felt experience and performed presentation may become genuinely difficult to access. The person is not suppressing feelings they clearly know they have.

In many cases, reliable access to the signals themselves has narrowed, because the mechanisms that would ordinarily carry signal into felt experience have been occupied, across time, by the sustained work of ensuring that signal does not reach expression.

This is what multiple first-person accounts describe, and what the empirical literature on camouflaging and identity also confirms. Pearson and Rose (2021), in a qualitative study of autistic adults' experiences of masking, found that sustained camouflaging was associated not only with exhaustion but with identity confusion and difficulty distinguishing authentic preferences and responses from performed ones.

This is not psychological fragility. It is a predictable consequence of sustained bodily labor occupying the mechanisms of self-reading.

Camouflage, Burnout, and the Invisible Pathway to Exhaustion

Section 8.6 identified two pathways to exhaustion: the visible pathway through repeated overload events with incomplete recovery, and the less visible pathway through sustained controlled activation that neither discharges nor fully resolves. Within this account, camouflage is proposed as a major contributor to that less visible pathway.

What is often described as autistic burnout can be understood, in this frame, as cumulative non-closure under sustained adaptive control — not overt escalation, but sustained controlled activation across time, closure deferred, the guard remaining engaged even when behavior appears compliant or calm. Raymaker et al. (2020) found that participants consistently described its onset as following extended periods of effortful social functioning and masking rather than acute overload events.

Burnout emerges as the endpoint of sustained adaptive labor rather than the aftermath of crisis. One participant described it as "the result of spending so much energy on trying to appear normal that there's nothing left for anything else — and it catches up with you all at once" (Raymaker et al., 2020).

Like other protective forms, camouflage supports functioning while increasing cumulative demand. It preserves outward presence while narrowing access to felt experience, deep rest, and spontaneous expression. Outward functioning does not necessarily reflect internal access. Because the costs are internal and functioning is preserved, the environments generating the demand — social, institutional, relational — receive no signal to reduce it. Camouflage is therefore self-concealing: it maintains the surface that prevents the conditions producing it from being recognized and changed.

Paired Scenario — Camouflage and the Hidden Cost of Outward Organization

It is a full school day. The same environment. The same sequence of demands. Both children are present and participating throughout.

For the first child, participation tracks available resources. When the task is within reach, the child engages. When activation rises, the expression of that rise is available — a pause, a request, a visible moment of effort. Output reflects the underlying state. The system's surface and interior are approximately aligned.

For the second child, participation is sustained from a system operating under a different condition. The environment requires a particular form of presence — engagement that meets expectations, expression that matches norms, responses produced at a pace and in a register the social context assumes. The system has learned that visible expressions of its internal state tend to generate responses that increase concurrent demand rather than reduce it.

A continuous monitoring process is recruited. Spontaneous expression is modulated. Facial configuration is adjusted. Timing is calibrated against what the room expects. This process runs in parallel with the task — not instead of the task, but alongside it, throughout the day.

Outward functioning is maintained. From the outside, this child appears regulated, engaged, appropriately present.

Internally, the monitoring process continues to draw on the regulatory resources available. Activation does not resolve because the guard remains engaged. Closure is deferred — each arc redirected inward rather than completed. What accumulates is not the visible cost of a threshold crossing but the invisible cost of sustained controlled activation across eight hours.

Both children leave the building. The day was the same.

The surface showed nothing. The floor has been rising since morning.

The Developmental Dimension

The developmental timeline along which camouflage consolidates varies substantially across individuals. Research on gender-differentiated camouflaging trajectories suggests that autistic girls and gender-minority individuals often develop camouflage earlier, more extensively, and with greater proficiency than autistic boys — a pattern Hull et al. (2020) associated with earlier and more intensive social-performance demands placed on female-presenting children, and one that correlates with substantially higher rates of missed and delayed diagnosis in these groups.

Lai et al. (2017) similarly found that autistic women showed higher camouflaging scores than autistic men even when controlling for autistic trait severity, suggesting that camouflage is shaped not only by individual regulatory history but by social environments that differentially reward or penalize its presence.

This does not mean camouflage is absent or less costly in autistic people outside these demographic pathways. It means that the conditions promoting its earlier development — environments that penalize visible autistic characteristics and reward social conformity — may operate with particular intensity in some developmental contexts, and that understanding camouflage requires attention to those conditions rather than assuming a uniform timeline.

Conditions for Recovery

When environments begin to permit authentic regulation — when expression no longer increases threat and reduced performance no longer carries relational loss — energetic expenditure may gradually decrease. Recovery becomes more possible. Regulatory margin may widen, not through social-skills training, but through reduction of chronic containment.

This shift cannot be produced by instruction or encouragement alone. The system does not lower camouflage because it has been told it is safe. It lowers when accumulated bodily evidence — repeated experiences in which expression did not increase load, presence did not require performance, and authenticity genuinely cost less than containment — reshapes the anticipatory organization camouflage was built to manage. This process unfolds at the pace of physiological learning, which is the pace of internal time. It cannot be accelerated by demand, but it can be supported by sustained conditions that make camouflage genuinely less necessary.

Closing — When Protection Was the Best Option

This chapter has traced how the body learns to safeguard itself under persistent demand and limited recovery. Responses often interpreted as problems — rigidity, silence, control, withdrawal, or excessive adaptation — emerge here for what they are: the most physiologically viable solutions available to a system organizing itself to continue.

Protective organization can sustain continuation under conditions the system cannot yet resolve. It reflects the system's response to conditions where recovery has not been sufficient to restore baseline — not dysfunction, but the logic of a system adapting to what sustained constraint required of it.

In the absence of reliably restored safety, the nervous system shifts. It narrows exposure, sustains alertness, and over time refines its protective patterns until they become increasingly difficult to see. This progression is one of the most viable ways of continuing without collapse when the conditions recovery requires are not reliably available.

Open protection and camouflage follow the same bodily logic. Both reduce physiological cost in contexts where showing, feeling, or responding carried excessive consequence. Both preserve viability through different pathways.

One discharges outward. The other contains inward. Each reflects the system preserving itself with the means available under the conditions it lived.

Protection, however, redistributes how resources are allocated. Sustained vigilance or controlled expression directs energy toward containment, narrowing internal access and limiting relational engagement. The system may continue to function, but with less space to feel, to recover, or to be without effort. In camouflage specifically, this narrowing extends to self-knowledge — to the reliability of the internal reading that would ordinarily allow a person to know their own states, preferences, and responses apart from what performance requires them to show.

This is physiological learning: how the body organized under conditions that made continuing this way necessary. Reading these responses from the body returns attention to the conditions that made them necessary and to the consequences they carry for being with others.

Ethical action is oriented toward conditions rather than performance — toward what the environment must provide for protection to become less necessary, rather than toward what the system can be trained to produce while protection remains in place.

How this translates into specific clinical environments — what it asks of practitioners in concrete settings and moments of accompaniment — is the subject the Clinical Workbook takes up directly.

The next chapter does not propose eliminating these configurations or demanding openness. It asks what unfolds when protection becomes the starting point of relationship — and under what conditions the system no longer has to carry this task alone or indefinitely.

Not to ask for more adaptation, but to understand what makes it possible for a body to share presence without losing itself.

PART IV — Relationship, Co-Regulation, and the Conditions for Reorganization

Part III traced what the body does when recovery is consistently unavailable: it protects itself. Through anticipatory alertness, environmental control, and finally through camouflage — the most refined and least visible form of protective organization — the system learns to sustain outward functioning while carrying the cost inward, preserving presence at the expense of access. The progression is not pathological. It is the physiological logic of a system that had to find a way to remain.

Part IV asks under what conditions these configurations may begin to reorganize.

Within this framework, reorganization is not expected to occur primarily through technique, insight, or reasoning that the current context is safe. The anticipatory architecture built by Chapter 10 does not update through argument.

The body does not lower its muscle tone because it has been told to. Heart rate does not slow because the accompanying person has explained that conditions are different now. Breath does not deepen because safety has been verbally promised. Updating occurs through experience — through the accumulated bodily evidence, provided across sufficient repetition, that different conditions are possible. Those conditions are often relational, but they are not only relational. Predictability where unpredictability had been expected, recovery where continuous demand had been anticipated, manageable sensory conditions where overload had been routine, and relationships that do not require protection all provide evidence that the organization built under prior conditions is no longer fully required. Part IV traces the specific conditions under which that reorganization becomes physiologically possible.

There is, however, a difficulty that precedes the account of reversal and must be named.

Part III established that the success of camouflage conceals its generating conditions: outward functioning is preserved, which means no visible signal is produced that would prompt a different relational offer.

Within this framework, the person most in need of reparative relationship may, structurally, be among the least likely to appear to need it. This is the initiation problem.

Visible distress is not a reliable signal for when a reparative relational field is needed — because the architecture of protection is organized to suppress precisely that signal. The physiological parameters traced across the preceding chapters provide a different basis for reading: not what the person in protection is able to show, but what the system's regulatory organization makes visible to a practitioner attending at the level at which this book has been working.

The chapter that follows examines what happens when protection meets a relational field that does not require it — and what accumulates, slowly, when that meeting repeats.

Chapter 11 — Relationship as Regulatory Condition

"I actually have lots of empathy. I'm just not really good at showing it. I am bursting inside with every single emotion one feels at all times." — Ethan Lisi, 2020

"The only time I felt like I didn't have to perform was with one particular therapist, and even then, it took about a year before I believed it was actually safe not to. The body takes time. It doesn't believe you straight away." — Autistic adult, participant in Crompton et al. (2022)

Opening — What Reverses Protection

This chapter is about the physiology of that reversal. It is a chapter about what a nervous system requires from a relational field in order for its protective organization to become, incrementally, less necessary — and what the accompanying person's own physiological state must provide in order to constitute that field.

Chapter 10 closed with a question: what makes it possible for a body to share presence without losing itself?

This chapter does not introduce a model of relationship. It describes the physiological conditions under which recovery and reorganization may occur when one nervous system is present with another.

The preceding chapters have built the conditions for answering it precisely. A system organized around anticipatory protection — whose floor has risen through accumulated residue, whose regulatory margin has narrowed across repeated unaccompanied activation, whose camouflage has been sustaining outward functioning at the cost of inward access — does not open in response to explanation, encouragement, or the reasoning that the current context is different from the ones that shaped the current traces.

The traces described here are not primarily narrative and may not be revised through narrative alone. What shaped them was the repeated physiological experience of what relational conditions provided or did not — accumulated across repetition until anticipation had calibrated around those outcomes. What may contribute to their modification is not explanation, but the re-encounter of those conditions under different outcomes: when remaining open does not produce the cost that prior openness produced, when the arc can complete rather than being carried alone, and when return occurs without urgency pressing it beyond what internal time allows.

The chapter moves in four steps. First, it establishes why emotion requires relationship to complete when intensity exceeds available margin. Second, it traces how steadiness in another nervous system modulates the activation arc in real time.

Third, it specifies the limits of co-regulatory approach — distinguishing the acute state in which approach adds to concurrent demand from the consolidated protective withdrawal that calls for a different response entirely. Fourth, it traces what reparative relationship provides over time: the specific conditions whose accumulated repetition allows anticipatory architecture to update. This sequencing matters: the limits of approach come before the account of reparative relationship, because understanding what does not work — and why — is necessary for understanding the precision that what does work requires.

11.1 — Why Emotion Requires Relationship

The same regulatory dynamics now appear in the relational domain. *Relationship*, as used here, does not refer to emotional support or interpersonal meaning, but to the physiological conditions created by the presence of another nervous system — and specifically to whether those conditions reduce or add to the concurrent demand the system is already managing. As established in Chapter 4, co-regulation is the direct physiological influence of one nervous system's state on another: a stable system in proximity supports recovery through the presence of regulatory input, not only through the absence of demand (Feldman, 2007; Porges, 2011).

Chapter 9 adds that when activation arcs complete without adequate relational conditions, residue is carried forward, baseline elevates, and anticipation recalibrates around the cost of what was not restored.

Throughout this chapter, "relational field" extends the earlier use of "*regulatory field*" to name the shared physiological space between two nervous systems in proximity. The term refers to the same continuous, co-varying state — now read as a dyadic rather than an individual organization. The principles governing it are the same; what changes is the unit of observation.

When an emotional arc rises beyond available margin — when intensity exceeds what the system can contain and complete through its own resources alone — the conditions for completion become relational before they become individual. This reflects the physiology of a social species: nervous systems regulate in the presence of other nervous systems, and the signals provided by a calm, regulated other contribute genuine physiological input to autonomic recovery rather than merely offering psychological reassurance (Coan & Sbarra, 2015; Feldman et al., 2011).

Completion, as used throughout this chapter, refers to the resolution of the activation arc through recovery processes that return the system toward baseline — not to emotional expression or cathartic release, but to the physiological descent that restores regulatory margin.

The requirement described here is physiological: when intensity exceeds available margin, the completion of the activation arc may depend on conditions that cannot be generated within the system alone.

Social baseline theory, developed by Coan and colleagues, provides a useful reference model for this account. The theory proposes that the human nervous system treats social proximity as part of its default regulatory environment, such that the presence of a safe other reduces the predicted effort required to manage a situation and effectively expands available margin (Coan & Sbarra, 2015). This framing requires an immediate qualification, however, because the default social baseline prediction — that proximity reduces regulatory effort — is precisely what the developmental history described in Part III may have revised.

For a system that has learned, through accumulated trace, that relational channels do not reliably reduce load, social proximity is no longer anticipated as regulatory support; it may be anticipated as an added source of concurrent demand.

What §11.1 adds is the direction of influence that is often overlooked: external relational conditions can change internal regulatory capacity, including available regulatory margin. This influence is not unidirectional. Each person's state becomes part of the other's regulatory environment, even when one person is currently providing the more stable input. Steadiness, then, is not merely a response to the person's state; it is also an input that can change what becomes possible within that state.

Social baseline theory describes a starting condition. Reparative relationship, over time, works by creating conditions under which that starting condition can be rebuilt — gradually, through the same mechanism of trace accumulation that revised it in the first place.

In many autistic profiles — where neuroceptive evaluation may operate with greater immediacy and less attenuation, and where the fine-threshold calibration described in Chapter 6 applies as much to relational signals as to sensory ones — the physiological registration of another's state may occur with particular directness. Small shifts in tone, timing, rhythm, or posture can widen or narrow regulatory margin before conscious interpretation has begun.

The system responds first to physiological state rather than stated intention. Breath rhythm, postural ease, and the unhurried quality of presence are often registered before language is, and they shape whether the activation arc finds conditions for completion or must continue to be carried alone.

The working orientation for this chapter is therefore precise: emotion requires a body in which to be felt, and often requires relationship to reorganize when intensity exceeds available margin. This is not because the autistic nervous system is incapable of self-regulation, but because the presence or absence of usable relational conditions materially changes what completion requires.

The sequences that follow are illustrative reconstructions. They show how the framework's constructs might apply moment to moment; they do not represent direct measurement of another person's internal physiological state.

Micro-Sequence — What Relational Availability Feels Like from Inside

A meeting has run forty minutes past its expected end.

The body is carrying what that cost has left. The system registers: someone is still present.
Not managing. Not directing. Simply available.

The room is quieter than the previous hour made it. Nothing is being asked.
Somewhere in the interval, something shifts.

Not resolve. Not calm. Not completion. Only: a slightly less urgent need to hold what is being held.
The other person's state is doing something. Not loudly. Not demonstrably.
Simply: the field is different from what it would be if they were not there.

Not changed feeling, but changed conditions. The arc is the same arc. The margin available to it is not.

This distinction — between changed feeling and changed conditions — is consequential. It prevents the most common misapplication of the co-regulatory account: the idea that the accompanying person's role is to make the other person feel better. The role is more precise and more demanding than that.

It is to hold, in one's own nervous system, a state that changes the physiological field of the interaction — and to hold it long enough, and with enough consistency, for the other system to register that the conditions have shifted. The feeling, if it changes, follows that registration. It is not produced by it directly.

This matters especially in autistic profiles where the body's response to relational presence may be more immediate and more physiologically direct than the surrounding culture's expectations prepare those in relation to them for. The person may not be able to name what they are registering. The registration precedes the vocabulary. This account holds that the body is responding to state, and that it does so before language has organized around the response.

11.2 — Steadiness as Regulatory Input

The presence of another person alters, moment by moment, the physiological conditions within which an activation arc proceeds. This is not influence through meaning or interpretation. Within this account, it constitutes a regulatory contribution of one nervous system's state to another, operating through the channels Part I Chapter 4 described: vocal prosody, breath rhythm, postural organization, interactional timing, and pacing of response.

Emotion does not reorganize in isolation. It reorganizes in relation.

The same organism can move through strong emotion in markedly different ways depending on relational conditions — not because the emotion itself differs, but because the conditions available for its completion differ. The arc rises as it rises; what changes is the field within which the descent occurs.

The vignette that follows is an illustrative reconstruction. It shows how the framework's constructs might apply within a relational scene; it does not represent direct measurement of either person's internal physiological state.

Conceptual Vignette A — The Arc in a Relational Field

A child moves into intense activation. Something has shifted — a transition, an unexpected change, an experience that arrived faster than integration could organize around it. The arc has crossed threshold and is carrying momentum. From the outside: distress, contraction, the body closing around what it cannot yet process.

One accompanying adult responds by moving closer, speaking faster, increasing the relational intensity of the moment — more words, more expression, more urgency. The child's system registers higher incoming signal load, faster prediction demands, and increased concurrent demand from the relational field. The arc does not settle. It may escalate.

A second accompanying adult remains nearby without pressing closer. Posture settles. Movements become deliberate and unhurried. When the adult speaks, the voice is lower and slower than the moment seems to invite. Nothing in the adult's state communicates urgency.

The child's system registers the difference: the relational field has shifted from adding load to reducing it. Muscle tone softens by degrees. Breath lengthens slightly. The arc does not collapse; it continues its course, but now within a field that is not driving it faster or higher. When descent begins, it proceeds with less resistance.

The arc was not shortened. The emotion was not redirected. What changed was the physiological field in which it unfolded.

The accompanying person is not controlling the child's emotion. They are providing the conditions — steady state, reduced demand, adequate time, non-urgency — within which the arc can complete through its own physiology.

Accompanied emotion does not disappear. It stabilizes. Experience can remain longer within regulatory range, move through rising and falling activation while coherence holds, and recover more readily toward restored baseline. The intensity is real and the peak is still reached. What changes is that descent occurs within a relational field that is providing rather than consuming what recovery requires. As stability consolidates, access to language, reflection, and exchange may expand. That expansion follows physiological integration; it does not initiate it.

A central principle the framework proposes follows from this account: performed calm may often register as such. The incongruence between verbal and physiological signal can itself constitute a prediction failure, adding to the concurrent demand of the interaction rather than reducing it. What the system responds to is physiological state, not stated intention — which means that the regulation required for genuine co-regulatory presence begins with one's own state. Not as a performance of calm, but as the actual physiological condition from which the interaction proceeds.

What Steadiness Requires of the Accompanying Person

It is equally important to specify what steadiness requires — because the most common failure is not the absence of care, but the presence of urgency carried inside apparent calm.

Steadiness, as the framework understands it, is not a behavioral style.

It is a physiological condition. A body carrying high load does not produce the same regulatory input as a body with available margin. Within this account, that difference constitutes a regulatory contribution to the system receiving it, because what operates between them is physiological state, not verbal content.

The person providing co-regulation must first assess their own regulatory state. Not to perform stability they do not have, but to honestly determine whether the conditions they are about to offer match the conditions they are intending to provide. An accompanying person carrying significant concurrent demand is not a neutral presence. Their state is information. The question is whether it is the information that supports recovery or the information that adds to the load the recovering system is already managing.

Co-regulation does not require perfection. What it means is that self-monitoring precedes other-monitoring. The first question in any co-regulatory moment is: what is my own state, and what is it contributing to this field right now? That question, asked honestly, is already a regulatory act.

Inconsistency and Reparative Limits

The response that works for most people may be the costliest available for this one, in this moment.

Reparative relationship does not require perfection. The accompanying nervous system will experience its own activation, distraction, and regulatory challenges. When the accompanying person's state becomes demanding or dysregulated, the system may temporarily return to protective organization.

This is not regression but appropriate adaptation to current conditions. Recovery of reparative possibility depends on the accompanying person's capacity to return to regulatory availability, not on sustaining it without interruption.

11.3 — The Limits of Co-Regulatory Approach

A necessary precision must be stated before the account of reparative relationship, because it defines what that relationship must actually provide. Two distinct forms of unavailability require recognition here, because they call for different responses.

The first is *acute*: a system that has been carried significantly above its functional access threshold, so that the signals associated with relational approach — touch, voice, proximity, facial expression — add to concurrent demand rather than reducing it.

The second is *chronic:* a system organized around the kind of protective withdrawal described in Part III, in which the learned expectation that relational channels increase load is the organizing condition itself, not a product of the current activation state.

These call for different relational responses: the first calls for reduced input and the passage of internal time; the second calls for the patient accumulation of reparative traces, which is the subject of §11.5.

In the acute state, the problem is straightforward. Touch, voice, and proximity all carry sensory concurrent demand. A face registering care still requires interpretation. A soothing voice still generates prediction updating. An outstretched hand still introduces the need to decide whether to receive it. At sufficient activation, these costs may exceed the regulatory benefit the approach is intended to provide.

The system in this state is not rejecting support. It is communicating, through its physiological response to approach, that the channels being offered are not currently receivable without additional cost. Reading this accurately is what the framework specifically asks for here: distinguishing withdrawal from rejection, and recognizing that increasing relational approach in response to visible distress may, for a system already at or above its functional access threshold, produce the opposite of the intended effect.

In some autistic profiles, this threshold may be reached more readily when conventional forms of relational approach themselves add processing demand. The point at which relational approach shifts from reducing concurrent demand to adding to it may therefore arrive earlier in the activation arc—not because the autistic system is less capable of co-regulation, but because the channels through which co-regulation is typically offered are not universally low-cost. Evidence from specific channels is consistent with this possibility: direct gaze, for example, has been associated with heightened subcortical activation in autistic participants under conditions of constrained eye contact (Hadjikhani et al., 2017). This finding does not establish that co-regulatory signals are generally more costly in autism; within this framework, it illustrates why the regulatory cost of a relational signal cannot be assumed from its conventional social meaning.

In the chronic state, what is needed is not more relational input but the sustained provision of conditions: quiet, predictability, reduced concurrent demand, and the passage of internal time at the system's own pace. The duration and intensity of these states do not necessarily correspond to the apparent severity of the precipitating event.

They follow internal processes moving according to their own rhythm of reorganization — the same internal time that governs integration and closure throughout this book. Regulation here does not occur through increased approach. It occurs through conditions that allow the system to use its own resources for recovery without adding to the demand those resources are already trying to clear.

The stance that supports recovery in both states is continuous with the chapter's broader account of support — reduced concurrent demand, predictability, adequate time, genuine safety — but delivered in the specific form each state requires. Presence becomes non-urgency. Support becomes the background reliability of an environment that is not adding demand, rather than active co-regulatory engagement. The accompanying person remains available without pressing. The field remains open without being filled.

The Same Limits in Family and Educational Contexts

The dynamic described above — in which well-intentioned co-regulatory approach increases rather than reduces concurrent demand — appears with particular frequency in family contexts and in educational environments, where the people most motivated to help are also most likely to misread withdrawal as rejection and respond with increased relational approach.

A parent who moves closer as a child enters overload, who speaks more and escalates the relational intensity of the moment in proportion to their own distress, is not providing the response the state requires at that moment. The issue is not a lack of care, but a mismatch between the response and the system's current receivability. What is needed is co-regulatory presence that matches the system's state, rather than presence that exceeds it.

In educational settings, the same mismatch appears in a different register. A teacher who responds to a student's withdrawal by increasing verbal engagement — asking what's wrong, offering reassurance — is following exactly the relational script that works for many students. For an autistic student already at or above threshold, it may be the most costly thing available in the moment. The words are processed as additional concurrent demand arriving precisely when the system has the least capacity to absorb it.

Reading this accurately in real time requires a reorientation that runs counter to trained instincts: when approach is not working, the skill is not to find better approach, but to provide withdrawal that is not abandonment. To step back, reduce input, remain visibly available without pressing, and allow internal time to do what urgency cannot. That tolerance for not acting is itself a regulatory resource.

A practical question this chapter raises but does not fully resolve: the framework specifies that a relational presence can function either as load or as restoration, but it does not provide a fully reliable in-the-moment signal for which is occurring. The observable indicators — the system's response to approach, changes in muscle tone and breath, whether activation rises or settles — require a level of attunement to physiological state that this framework describes but cannot teach in the abstract.

What the framework can offer is the question to carry: is my presence, right now, reducing concurrent demand or adding to it? That question, asked honestly and read from the body rather than from intention, is the closest available substitute for a signal the theory has not yet specified precisely. This is a genuine gap, not a pedagogical one. A future empirical contribution to this framework would be the specification of observable physiological markers that allow the distinction between load and restoration to be made in real time. The candidate observations described in Table 1 — HRV trajectory and heart rate recovery slope — provide the closest available physiological anchors for this reading, even when formal measurement is not possible. Attending to them trains the attunement the question requires.

11.4 — The Physiology of Reparative Traces

In simpler terms: each time an activation arc completes in a different relational condition than the one that built the protective architecture, it may leave a small physiological record. Enough such records, accumulated over time, can begin to revise what the system predicts relational presence will cost.

What §11.5 calls a reparative trace is proposed here as a physiological record left by an activation arc that completed differently from those that built the current protective architecture — differently in the conditions available during its descent, and therefore differently in the physiological cost associated with remaining open in that relational context.

The account that follows is a theoretical extension of the framework's predictive architecture applied to relational learning. While grounded in adjacent empirical literatures — extinction learning, latent inhibition, autonomic indices of recovery — the specific formulation advanced here remains inferential and awaits direct empirical testing. This framework remains an interpretive organization of existing knowledge and requires empirical testing to determine its measurable correlates and predictive validity.

In the terms Part I established, the framework models a completed reparative sequence as an activation arc whose descent occurs within a relational field that reduces concurrent demand rather than adding to it. If the proposed process has measurable autonomic correlates, candidate indicators could include patterns of parasympathetic re-engagement, high-frequency HRV trajectory, heart-rate recovery, electrodermal activity, and changes in muscular tone during and following relational activation. The discrepancy between predicted and experienced recovery cost is not itself the trace; within the RNS hypothesis, it is one condition from which a fractional revision of relational prediction may arise. Extinction and prediction-error models provide an explanatory analogue for this possibility rather than a demonstrated mechanism of reparative relationship.

The accumulation of such traces is proposed here as a physiological process through which reparative relationship may alter subsequent regulatory organization. Within the proposed model, revision is expected to recruit learning processes related to those involved in the original protective organization, which may help account for its gradual pace.

The pace of reparative accumulation — which may feel slow to the accompanying person, and which does not produce visible inflection points — is not evidence of treatment failure or relational inadequacy. It reflects the pace the framework proposes as consistent with extinction-based learning principles — a process that, on this model, builds competing traces rather than erasing prior ones, and which may require substantial repetition, consistent with the persistence of previously learned protective associations. Expecting rapid revision may therefore be inconsistent with the gradual updating process proposed here.

That asymmetry — between the pace of protective learning and the pace of its revision — is documented in the extinction learning and latent inhibition literatures. Protective associations formed under conditions of developmental necessity, often early and with high repetition, resist revision by new experience more strongly than associations formed later or with less repetition.

Extinction does not erase the original learning; it builds competing traces alongside it, such that the original prediction is still available — still capable of being activated when conditions resemble those that formed it — but is no longer the only prediction the system can generate.

What reparative relationship builds, over sufficient repetition, is an expanded predictive repertoire: not the absence of protective anticipation, but the presence of an alternative that the system can reach when present conditions resemble those in which reparative sequences have previously completed.

Put simply: the RNS hypothesis proposes that repeated relational experiences that cost less than anticipated may gradually revise what the body expects similar relational conditions to cost.

Candidate physiological indicators of the proposed update process would be expected to change gradually and non-linearly rather than provide a discrete signature of a reparative trace. Early in a reparative relationship, autonomic patterns may differ little from prior sessions. Across repeated sequences, the RNS hypothesis predicts that relational approach under comparable conditions could become associated with lower anticipatory cost, later protective mobilization, more complete recovery, and broader expressive range. These are predictions of the framework, not established markers of an underlying reparative-trace mechanism.

These predictions are, in principle, testable. Longitudinal work could examine HRV trajectories, heart-rate recovery slopes, electrodermal activity, and other contextually appropriate autonomic measures across repeated relational sequences. Such measures would not directly measure a reparative trace; they would test whether the physiological patterns predicted by the RNS formulation occur as proposed.

11.5 — Reparative Relationship: Living Something Different

At the point where protection has become anticipatory — where the system is organized less around the current relational context than around what prior contexts physiologically demonstrated — what changes that organization is not explanation of how the current context differs. Within this framework, these traces are treated as physiological, and their revision is therefore expected to depend on lived regulatory experience rather than conceptual understanding alone.

The system most in need of these sequences is, structurally, least likely to signal that need through visible distress — because camouflage has organized precisely around sustaining outward functioning. The relational field is therefore available before visible signal — held open through reading of physiological state rather than behavioral expression.

The initiation is not the system's responsibility.

This is the precise sense in which relationship can become reparative: not through the reasoning that things are different now, but through repeated, concrete, bodily experience in which the activation arc unfolds under conditions different from those that built the current traces.

The concept has roots in both developmental and clinical literatures.

The corrective emotional experience, first described by Alexander and French (1946) and elaborated across later relational and attachment traditions, names the observation that what revises anticipatory relational organization is not insight about the past but lived experience in the present, sustained across sufficient repetition for implicit learning to update its anticipatory architecture (Siegel, 2012; Schore, 2012). The framework developed in this book proposes an explanatory model: each sequence in which an activation arc completes under different conditions may leave a trace. As such traces accumulate, they introduce variation into anticipatory architecture. Over time, that variation can shift the prediction the system makes about what relational conditions provide.

What updates through reparative repetition are the somatic predictions that guide anticipatory organization. Each encounter in which authentic expression is met with steady rather than demanding response creates a physiological record — a trace laid down in the same predictive architecture that built the original protection. In concrete terms: when another person consistently receives distress without escalating their own activation, who waits without filling the silence, who returns reliably after the person retreats — these specific, repeated experiences are the data from which the nervous system updates its prediction about what relational presence costs.

As such traces accumulate, the system's default prediction about relational cost may begin to shift. This updating occurs through dynamics continuous with those described in Part I: prediction updating, accumulation across repeated cycles, and shifts in threshold proximity. The difference is that these dynamics are now being applied to relational rather than sensory or cognitive organization.

Reparative relationship can be described through three recurring relational sequences.

Activation rises without collapse. The traces that built anticipatory protection often formed through repeated arcs that escalated beyond what available margin could contain, leaving residue from which the next arc began. In the reparative sequence, activation still rises, but it rises within a relational field whose state reduces concurrent demand rather than adding to it.

What changes is not the reality of the emotion, but the margin available for the system to remain within regulatory range while the arc proceeds. The accompanying person's steadiness does not suppress the emotion. It provides the conditions — reduced load, predictability, non-urgency — that allow the arc to remain containable rather than escalating beyond available organization.

Retreat occurs without disconnection. Protective withdrawal often developed in contexts where retreat itself increased load: urgency from others, demand for immediate return, or the relational cost of stepping back. In the reparative sequence, retreat does not sever the relational field.

The accompanying person remains present without pressing, without filling the silence with demand, and without reading withdrawal as rejection. What the interval needs is the absence of pressure to end it — no urgency filling the silence, no signal that the retreat has continued long enough, no re-engagement before internal time has completed its work.

The system can then use the interval for recovery at the pace internal time requires, without losing the connection to which it will return. Retreat becomes a viable regulatory option rather than a relational penalty.

Return happens without pressure. The pace of recovery is governed by internal time, not by the convenience of the surrounding environment. In the reparative sequence, the accompanying person does not set the timeline for re-engagement. They remain available. Return occurs when the system has completed enough internal reorganization for re-entry to be physiologically possible. Across repeated experiences of this kind, anticipatory prediction begins to change: return is no longer expected to be demanded before it can actually occur.

These three relational sequences are concrete expressions of the recovery conditions established in Part I Chapter 4 — reduced concurrent demand, predictability, adequate time, and relational safety — now provided within a relationship often enough, and reliably enough, for anticipatory architecture to begin revising its predictions. Reparative relationship is not a therapeutic technique. It is any relationship that consistently provides these conditions in the domain of emotional activation, allowing the system to accumulate traces of a different relational experience.

Conceptual Vignette B — What Accumulates

The first time it happens, the system barely registers it. A small activation arc — nothing like the experiences that built the current traces. The other person remains steady. The arc rises. Return begins. The interval is not rushed.

One sequence. A slight trace.

Within a handful of sequences, something is fractionally different. No narrative has formed around it. The body simply registers that this arc did not require the expenditure it predicted. The prediction is not revised. But the discrepancy between prediction and experience is proposed to register physiologically.

After enough repetitions that the body has some evidence, the system is entering a relational field about which it knows something. Not certainty, but no longer nothing. The preceding sequences have left traces — partial, variable, still insufficient to reorganize the larger architecture, but present.

After more still, something has shifted. Not as decision or conscious understanding. The system is beginning to organize anticipation around slightly different evidence of what this relational field provides. The arc rises more readily — not because the feeling is weaker, but because the predicted cost of allowing it to rise has begun to decrease.

The specific pace of this accumulation is not fixed. It depends on the depth of prior trace, the intensity of each reparative encounter, and the individual parameters of the nervous system.

What the sequence above illustrates is a direction, not a timeline.

Nothing about this is fast. Within the proposed model, reparative sequences are expected to accumulate across repeated encounters before protective predictions may begin to revise.

The pace of this process is governed by the same internal time that governs every other physiological timescale traced in this book. It cannot be accelerated by demand, by explanation, or by increasing the frequency of relational sequences beyond what recovery capacity can support. It unfolds at the pace of physiological learning: the pace at which new traces are added, and old predictions encounter enough contradictory evidence for revision to begin.

What the accompanying person provides, in this frame, is not treatment. It is the consistent maintenance of conditions that make the system's own learning possible.

The reorganization is the system's.

The conditions are the relationship's.

As these sequences accumulate, structural changes unfold gradually. Anticipation softens — not because prior traces are erased, but because new traces introduce enough variation that prediction is no longer uniformly organized around a single expected outcome.

Vigilance loosens. Expression becomes more possible without immediate protective override, not because regulation is being effortfully achieved, but because it is no longer being carried alone.

Access to one's own states may also become more reliable, as less energy is required for the self-monitoring and containment that earlier conditions demanded.

This is not recovery to a prior state. For many autistic nervous systems, there may be no such prior state to return to; protective organization may have developed early enough to precede the formation of the very traces now being revised.

What this process offers is not restoration but expansion: a wider activation range within which the arc remains containable, a more reliable channel between felt experience and available expression, a recovery speed that is no longer structurally impaired by continuous vigilance. Not the absence of protection, but protection that is less continuously required — and a broader lived sense that relationship can provide rather than consume what recovery requires.

Extended Sequence — The Accumulation Arc

In the first months, the system reads the accompanying person as it reads any new relational context: with fine-threshold calibration, without accumulated evidence, and with the predictive architecture that prior relational experience has built. Approach is registered. Small signals are tracked. The system is building its record.

There is no dramatic turning point. The accumulation is sub-narrative. What changes, across many encounters, is not that the person proves themselves. It is that the record is building. The system has enough evidence to construct an alternative prediction — not to replace the protective one, but to stand beside it.

After a year, something may be observably different — though the person whose system has been changing may be the last to notice, because the change has been gradual enough to remain below the threshold of narrative.

What has changed: the latency before protection activates in this specific relational context has lengthened. The cost of remaining open, in this presence, has decreased.

None of this was produced by insight or explanation.

It was produced by the consistent provision of conditions over sufficient time.

The relationship did not fix anything. It provided the ground from which the system's own processes could proceed differently.

This sequence illustrates the non-linearity the framework predicts. Progress in reparative relationship does not appear as a trend line. It appears as expansion within specific contexts, followed by apparent regression when conditions change, followed by gradual generalization as the reparative prediction becomes more robust.

Closing — The Body Returns in Relationship

Emotion does not complete in isolation. It can be felt within the organism — the arc rises, the intensity is real, the bodily process is genuine. But once intensity exceeds available margin, the conditions for completion become relational before they are individual.

With adequate support — offered through channels the system can receive without additional cost, at the pace internal time requires, from a field that is genuinely steady rather than performed — expression and protection no longer need to compete.

The arc can complete. What completes leaves a different trace from what was left incomplete: a bodily record of moving through intensity in the presence of another who remained steady, whose field reduced rather than added to concurrent demand, and who did not press return before internal time permitted it.

Relationship does not erase difficulty. It changes how experience is carried. The difference between an arc carried alone and an arc carried in accompaniment is also a difference in what the body comes to anticipate. Part III traced the protective configurations that develop when activation is repeatedly carried alone.

What develops through repeated accompaniment is a different

anticipatory architecture: one in which relational presence is increasingly predicted to reduce rather than increase load, and in which completion in relationship becomes physiologically possible.

This transformation does not occur through explanation, insight, or intention alone. It unfolds through lived repetition, at the pace of physiological learning, within relational conditions that allow the system's own processes to proceed. As sequences accumulate in which retreat does not sever connection and return is not rushed, anticipatory architecture begins to update — not through revision of what was, but through accumulation of what now is.

From this place, a foundational capacity becomes more available: remaining present as intensity shifts without immediate protective override. Not because protection has been eliminated, but because the relational conditions the system has been encountering have been providing rather than consuming what recovery requires.

Margin widens because relationship has been contributing to it.

What this looks like in a single moment is not dramatic: one person remains steady while another moves through something difficult, and neither the steadiness nor the difficulty is performed. The arc rises. The field holds. Descent occurs.

Something is left behind that was not there before — not a change in what the person can be asked to do, but a fractional change in what the body predicts this particular presence will cost.

What follows asks what it means to apply that reading in real time — in concrete contexts, in moments when a body is showing what it is carrying and a decision must be made about what to do next. Not which technique to apply, but what level of attention to bring, what state the system is in, what conditions the next moment requires, and what the accompanying person's own state makes available to provide.

The reading practice is not separate from the relational practice.

They are the same practice at different levels of resolution.

This framework offers one lens for understanding nervous system organization and autistic experience. Other approaches — developmental, phenomenological, social, political — illuminate dimensions of autism that this physiological account cannot address. The regulatory perspective developed here complements rather than replaces these other ways of knowing.

PART V — Clinical and Educational Implications

What changes when a practitioner can read the nervous system rather than only observe behavior.

The architecture has been traced. What follows asks what becomes visible when it can be read.

Parts I through IV traced the architecture of nervous system organization — how regulatory parameters interact, how autistic organization differs structurally rather than deficiently, how protective configurations develop when conditions require them, and how relational fields can provide or consume what recovery requires.

Part V asks what it changes in practice.

The answer is not a set of techniques. Reading the nervous system does not replace intervention approaches. It helps clarify the conditions under which any approach becomes possible. Techniques, observation tools, and applied practice structures belong to the companion volume, Reading the Nervous System: The Clinical Workbook, which translates this framework into concrete practice across clinical, classroom, and family contexts, with scaffolding, worked examples, and progressive skill-building sequences. What this volume provides — and what Part V traces — is the prior move: the observational reorientation that makes those tools coherent rather than procedural.

Many experienced practitioners already respond intuitively to shifts in regulatory state. They slow pacing when activation rises, reduce complexity when access narrows, recognize when participation is becoming physiologically costly, and adjust relational conditions before visible disruption fully emerges. Often, this occurs without a formal physiological language for what is being observed. The contribution of this framework is not the invention of a new clinical capacity, but the organization of an existing perceptual one: making the reading of nervous system conditions more explicit, more transmissible, and more available under pressure.

These adjustments are not directed only toward the person. They are often directed toward the conditions within which the person is organizing. Reading the nervous system therefore makes environmental influences more visible as well, allowing support to be directed not only toward the system itself, but toward the demands, predictability, recovery opportunities, and relational conditions shaping what the system must sustain.

A clinician or educator who has absorbed the preceding framework is no longer limited to reading behavior as output. They can begin to read the regulatory conditions that generate it — and in doing so, ask different questions, make different decisions, and offer different kinds of support. They may also begin to recognize that capacities often interpreted as stable traits — including curiosity, flexibility, reciprocity, exploratory engagement, or orientation toward novelty—can fluctuate with the regulatory conditions available within the system. As regulatory margin narrows, the range of experience the system can sustain often narrows with it. As access expands, these capacities may begin to re-emerge, not necessarily because the person has changed, but because more of the system has become physiologically available.

Chapter 12 traces that shift in observational practice and its implications for how support is conceptualized. It is not a clinical manual. It is a reading orientation: a description of what becomes visible once the framework is in use

— and what that visibility opens.

Chapter 12 — What Legibility Opens

Why Physiological Reading Changes What We See

"I had not seen it coming. I had not seen what his body had been carrying."
— María E. Reyes, M.S., CCC-SLP, clinical session notes

Reading the nervous system does not mean assigning fixed meanings to behavior or achieving certainty about internal states. It means learning to observe patterns across time, context, regulation, recovery, and changing conditions. The framework does not eliminate trial and error, because living systems remain dynamic, multivariable, and state-dependent. What it provides is orientation: a way of clarifying where to observe, what variables may matter, and how the system's organization shifts as conditions and relationships change.

Opening — What the Framework Makes Readable

A practitioner enters a session or a classroom and sees a child who will not begin the task. Or a student who was engaged yesterday and is unreachable today. Or an adult client who describes themselves as "fine" while every visible physiological indicator suggests otherwise. Behavioral observation registers these as events. The framework built across the preceding chapters makes them readable as states — as the outputs of regulatory conditions that are currently organizing what is and is not available.

That shift — from event to state, from output to condition — is what Part V traces. It does not specify what to do. It specifies what becomes visible, and why that visibility changes the questions a practitioner can ask.

12.1 — Why Behavior Alone Is Not Enough

The session was proceeding as expected. He was cooperative and engaged — present in the way that, from the outside, looks like readiness. Then, without visible transition, he began throwing materials to the floor. He climbed onto the table, screaming. I had not yet considered what his body might have been carrying into that room before a single demand was placed — the accumulated load the preceding hours may have required him to manage, the margin that may already have been reduced, the possibility that he had arrived closer to threshold before I began.

Behavior was what I could see. What I did not yet know how to consider was the regulatory field within which that behavior had emerged.

The limit is not accuracy but depth: it registers what occurs, but does not specify the conditions that generate it.

Two children who refuse to enter a room are producing the same behavioral output. One is carrying accumulated residue from a morning of continuous concurrent demand and has narrowed to the point where any new threshold-crossing event exceeds available margin. The other has encountered a specific sensory condition in that room — a particular acoustic quality, a lighting frequency, a smell — that constitutes a prediction mismatch steep enough to produce immediate protective reorganization. The behavior is identical. The generating conditions are not. The support that addresses one may not address the other.

Figure 23 maps this bidirectionality: the same output does not specify the generating condition, and the same condition does not produce the same output.

Figure 23 — Behavioral Output Does Not Specify Regulatory State

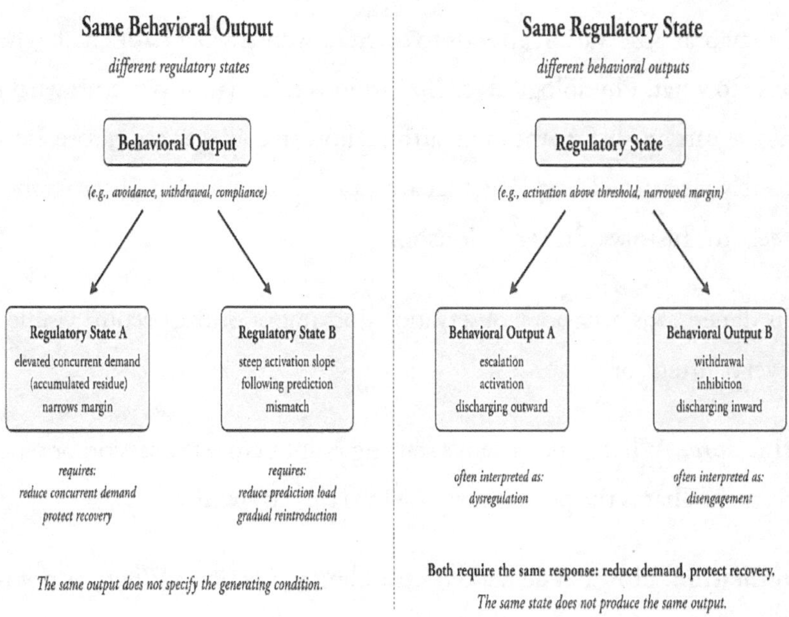

Figure 23 – Behavioral Output Does Not Specify Regulatory State

Left: the same behavioral output can emerge from different regulatory states — making behavior alone an insufficient basis for identifying what the system requires.

Right: the same regulatory state can produce different behavioral outputs — making behavioral difference an insufficient basis for inferring the underlying state.

This figure is not a model to be applied. It is a simplified reading of a process that is already occurring continuously.

Reading the Nervous System • María E. Reyes, M.S., CCC-SLP

This is the observational problem the framework corrects. Behavioral reading is indispensable. Physiological reading makes it interpretable.

12.2 — What This Framework Changes in Clinical Observation

The shift the framework produces is not additive — it is not a new checklist appended to existing observation. It is a reorientation of what the clinician or educator is tracking across time.

Behavioral observation attends to discrete events: what happened, when, in response to what. Physiological reading attends to dynamic patterns: what the system was carrying before the event arrived, how steeply activation rose, how completely it returned, and how that arc has been accumulating across the day, the week, the history of this relationship.

Six dimensions behavioral observation does not prioritize become visible from this level of attention:

Baseline state. What is the system's starting point before the session or school day begins? What is the person carrying into this context?

Accumulation. How has demand been adding across time? What has not fully recovered before the next demand arrived?

Threshold proximity. At what point in the session does the distance between current activation and threshold begin to close? What conditions are driving that closure?

Recovery quality. When the system has an opportunity to recover, does it return fully? Partially? Not at all? What does that trajectory suggest about current residue?

Environmental fit. Which features of the current setting are adding to concurrent demand without being recognized as demand — sensory conditions, predictability failures, relational ambiguity, transition frequency?

Internal physiological state. What is the body itself contributing to the current load? Unresolved internal physiological load is concurrent demand in itself — not a condition that modifies how the system responds, but load the system is already carrying before any external demand has arrived. Before environmental fit is evaluated: what is this body already carrying?

Attending to these dimensions does not require formal measurement. It requires a reorientation of attention — from "what is this person doing" to "what is the system carrying, and what does this context require of it." That shift is itself a clinical skill. It develops through practice, not through procedure.

It also requires something of the practitioner's own state. A clinician or educator operating at high concurrent demand themselves is less available to the fine-grained reading this asks for. The observational shift is not only outward. It requires the same quality of attention to one's own regulatory state that Chapter 11 identified as the precondition for genuine co-regulatory presence.

Within this framework, the effectiveness of any strategy, intervention, or therapeutic method depends in part on whether the nervous system has conditions it can organize as sufficiently safe. *Safety* is not a secondary feature of treatment and not a background condition beneath the work. It is itself part of the regulatory process through which access, participation, flexibility, learning, and relational availability become possible.

From this perspective, creating conditions of safety is not separate from intervention. It precedes and enables it. The practitioner is not only introducing techniques or demands into the environment — the practitioner is also shaping the regulatory conditions under which the nervous system organizes toward participation, protection, or defensive allocation.

Nervous systems reorganize through repeated conditions, not through instruction alone. The effectiveness of a technique depends not only on what is applied, but on whether the conditions present are compatible with the regulatory access that technique requires. An approach that works when the system has available margin may fail when margin has already narrowed — not because the approach is wrong, but because the conditions that make it possible have not been established first.

12.3 — Reinterpreting Common Clinical and Educational Assumptions

The shift in attention described above has direct consequences for how familiar observations are interpreted in practice. Looking beyond the output changes the question from what this person is doing to what conditions the system may be carrying that could be shaping the responses available right now.

The RNS reading offered below is one hypothesis among others that remain possible. Reading through this framework does not establish which explanation is correct. It adds a dimension the professional can examine by adjusting conditions, observing what changes, and remaining willing to revise the interpretation.

Clinical setting

Behavioral observation	Possible RNS hypothesis	Other explanations that remain possible
Behavior described as "noncompliance"	Protective reorganization: functional access to task performance may have narrowed below what the current demand requires	Fatigue, unaddressed pain, task misunderstanding, valid disagreement
Avoidance	Demand may exceed available margin: engagement may require more than the system can readily sustain under present conditions	Prior negative experience with the task, sensory aversion unrelated to current load, competing priority
Behavior described as "overreaction"	Steep activation slope combined with narrowed margin: the rise may be fast relative to the containment currently available	A proportionate response to a genuinely significant event; unaddressed pain or illness
Inattention	Concurrent demand saturation: processing capacity may already be allocated elsewhere, making it harder to absorb incoming signal without cost	Boredom or under-stimulation, sleep deprivation, difficulty understanding the task
Performance decline under load	State-dependent functional access: changes in regulatory conditions may be contributing to reduced access to previously demonstrated skills	The skill was never fully acquired; unclear instructions
Inconsistency	State-dependent accessibility: engagement may be more available under some regulatory conditions than others; observed inconsistency may reflect differences in access rather than differences in willingness	Differing task demands across contexts; motivation differences

Educational settings

Behavioral observation	Possible RNS hypothesis	Other explanations that remain possible
Task refusal	Demand may exceed available margin: the request may arrive at a point where compliance is difficult to sustain without protective reorganization	Valid disagreement, unclear expectations, an unaddressed physical need
Selective participation	State-dependent accessibility: engagement may be more available under some regulatory conditions than others; observed inconsistency may reflect differences in access rather than differences in willingness	Interest-based engagement — a normal variation, not necessarily physiological
Emotional outburst following minor trigger	Accumulated residue: the response may reflect what the system was already carrying, more than the size of this particular trigger	An unrelated stressor; a developmentally typical response
Apparent inability to generalize learned skills	Margin-dependent transfer: skills acquired under low-demand conditions may be harder to access when the learning environment itself constitutes significant concurrent demand	The skill was taught in a context-bound way; insufficient practice across contexts
Withdrawal or shutdown during transitions	Transition cost: each environmental shift requires prediction updating; cumulative transition load may exceed available margin without any single precipitating event standing out	Unclear expectations about the transition; an unrelated emotional response

In both settings, the reinterpretation does not change what was observed. It changes the explanatory frame — and therefore the question that follows. A behavioral frame asks: how do I change this response? A physiological frame asks: what conditions are generating this response, and which of those conditions am I in a position to alter?

12.4 — Why Support Must Target Conditions, Not Just Behavior

Changing behavior without changing the conditions that generate it produces, at best, a surface-level shift—and often increases the cost the system is already carrying.

When available regulatory margin is already limited, additional performance demands may consume remaining capacity rather than expand it.

Across repeated experiences of this kind, the system learns that this context reliably requires expenditure beyond available resources—and organizes accordingly. Anticipatory protection does not develop arbitrarily. It develops in response to exactly this pattern.

Sustainable change requires altering the conditions within which the system operates—not instead of addressing behavior, but as the prior move that makes behavioral change physiologically possible rather than physiologically costly. The four parameters identify where to look when conditions, rather than outputs, become the focus.

Because the system operates as a coordinated field, change does not occur by isolating parameters, but by altering the conditions under which they organize. Interventions that target one aspect in isolation risk redistributing load rather than reducing it.

The examples that follow are illustrative reconstructions. They show how observable patterns can be interpreted through the framework; they do not represent direct measurement of another person's internal physiological state. Terms such as concurrent demand, activation slope, regulatory margin, and recovery identify plausible dimensions for observation and inference, not internal conditions that can be known with certainty from behavior alone.

Concurrent Demand

> *Eight forty-five. He arrives at the classroom already carrying the morning — the unpredictable commute, the hallway noise, the signaled change in routine. By the time the first task is placed in front of him, his baseline is already near or past threshold. The task is the event that makes the load visible. The load has been accumulating since he left the house.*

The system cannot absorb what it is still managing. What reads as resistance to the task is the expression of an elevated baseline already in place before the session began. The demand that produced the visible response was not the task itself, but everything that preceded it. Reading concurrent demand requires asking not only what is being asked of the system now, but what the system was already carrying when the asking began.

Within this reading, the response did not begin here. The regulatory conditions shaping it were already in motion.

When concurrent demand is considered with greater contextual precision, the meaning of the moment can change. The task need not be assumed to be the sole source of the difficulty; it may be the point at which accumulated load becomes visible. The question shifts from what the person is refusing to what the system may still be managing. This reading invites attention to whether reducing demand and supporting recovery changes access to learning, participation, flexibility, or communication.

Co-regulatory support — not as behavioral reinforcement, but as physiological conditions that allow the system to return toward regulation — becomes the prior move. It does not reward the response. It addresses the conditions generating it. Only once sufficient margin is restored can new demand be placed — and when it is, it is placed into a system that can absorb it.

Environmental Modification

Environmental modification, within this framework, refers to changing the conditions under which the system is currently organizing. This includes the demands already active, the predictability already available, and the relational field already present.

If the system is already carrying more than it can absorb, the first intervention is not adding something new. It is changing the internal and external conditions that are shaping organization in the present moment.

Activation Slope

Ten fifteen. The schedule changes — one sentence, delivered neutrally. For her, the prediction organizing the morning is invalidated in a moment. Activation rises steeply. By ten seventeen, the slope is still ascending even though the announcement has already ended. The room has moved on. Her system has not. What reads as an overreaction to a minor change is the continued rise of a steep activation arc, still unfolding minutes after the trigger has passed.

Steep activation slope is not fragility or poor regulation. It is the rate at which this system mobilizes when expectation and incoming signal diverge — a feature of how the system is organizing under those conditions, not a choice. The magnitude of the response is not determined by the size of the change, but by how quickly activation recruits once prediction fails, and how much margin was available when that recruitment began. The delay between trigger and visible peak is not irrationality. It is the arc continuing toward its apex.

Within this reading, the event need not be understood as creating the response. It may reveal an activation arc already underway.

When a steep activation slope is a plausible reading, the moment of visible response need not be treated as the beginning of the process. The activation arc may have begun before the behavior appeared and may continue after the apparent trigger has passed. Pressing for re-engagement while activation appears to be rising can add demand to a system already in motion. A reasonable adjustment is therefore to test a steadier field — reduced input, maintained predictability, and time — and observe whether access changes as activation settles before reintroducing the change with preparation.

Regulatory Margin

One fifteen. She has been managing since nine — not struggling, managing. Producing expected outputs, sustaining participation, tracking the relational demands of a busy classroom. The margin from which those outputs have been produced has been narrowing since morning. At one fifteen, a peer makes a comment she would have absorbed easily at nine. She cannot absorb it now — not because the comment is different, but because threshold proximity has increased until almost no margin remains. What reads as a sudden reaction is the loss of the last available margin.

Regulatory margin is not a fixed property. It is the available range within which the system can absorb new demand without activation exceeding what it can sustain — and that range shifts across the day as demand accumulates and recovery either occurs or does not. Reading margin means reading the arc that produced the moment: what the morning required, what recovery was available between demands, and how much capacity remained when this event arrived. Within this reading, the afternoon event may have arrived when available regulatory margin was already substantially reduced.

The event may therefore be less informative than the regulatory conditions into which it arrived.

When narrowing margin is a plausible reading, the implication is not to address the event in isolation. One question is whether the system may have been operating at increasing cost across time, leaving less capacity to absorb ordinary demand. A clinically available test is to reduce demand and protect a genuine recovery interval before introducing further demand, then observe what changes. The aim is not to claim knowledge of an internal regulatory need, but to examine whether a different accounting of what the entire day has required better explains the pattern.

Recovery

Eleven thirty. The session ended twenty minutes ago. He worked hard — genuinely hard — and the work was good. The room is quiet now. He is sitting. From the outside, this reads as recovery. Internally, activation is still descending from a peak reached near the end of the session. The arc has not yet completed. When the next practitioner enters and the next session begins, he cannot meet it from where the first session ended. He can only meet it from where recovery has brought him — which is not yet back to baseline. What reads as reduced engagement is the carried-forward residue of an arc that was not allowed to close.

Recovery follows the system, not the schedule. The interval between demands is not neutral time — it is the window in which activation either resolves or carries forward as residue. When that window is shorter than the system's internal time requires, residue accumulates. The next demand is met from an elevated baseline. Across repeated cycles of abbreviated recovery, that baseline continues to rise, and each subsequent demand is met with less available margin than the one before.

The difficulty does not begin here. It is the continuation of an arc that did not complete.

When incomplete recovery is a plausible reading, the implication is structural: test whether protecting the interval rather than filling it changes what follows. Conditions that reduce demand may support physiological recovery, including parasympathetic re-engagement as one component of that broader process. This is not rest as reward. It is an opportunity to examine whether greater recovery before the next arc changes subsequent access. If it does, that pattern strengthens — but does not by itself prove — the inference that incomplete recovery was contributing.

12.5 — What Legibility Makes Clinically Visible

Understanding why something is happening is not the same as validating it, excusing it, or removing the need for limits. The framework describes the conditions under which responses become probable. Explaining the origin of a response does not reduce its impact or its consequences. It makes the response legible within the system's organization — and legibility is precisely what opens the possibility of effective intervention.

What it does is shift the question. Not: is this response acceptable?

But: what conditions made this response the most available one, and which of those conditions can be altered?

When that question is asked accurately, three things become more likely than behavioral management alone can reliably produce: disruption and aggression may become less probable as the conditions generating them change, learning may expand when margin is available to absorb it, and curiosity — which may depend on supportive physiological conditions — becomes more likely to appear. When regulatory conditions are misread as behavior, support targets the output rather than the system — and often increases the physiological cost it was meant to reduce.

When Disruption and Aggression Reduce

Within an RNS reading, disruptive and aggressive responses may reflect the system's most available response when concurrent demand has exceeded available margin and no other regulatory option remains accessible. Within this framework, these patterns should not be interpreted primarily as expressions of will, defiance, or character.

One RNS interpretation of the session described in §12.1 is that accumulated demand across hours had narrowed available margin before the final demand arrived, leaving the system without another regulatory option once that demand could not be absorbed. The disruption was not the problem. It was the signal. The problem had been building long before anything was visible at the behavioral surface.

The intervention that changes disruption is not limited to the moment it appears. It also addresses the conditions building toward it — across the hours, the days, and the relational history preceding it. The aim is not merely a surface-level shift, but a change in the conditions within which the response becomes more or less probable. These shifts are not immediate or uniform. They become more probable as conditions change consistently over time.

When Learning Expands

Learning depends on available processing resources. Within an RNS reading, regulatory margin provides one way of thinking about how much capacity may remain available for encoding, integration, and transfer under current conditions. When the system is already operating under high demand, access to those processes may become more constrained. This is why the same child who cannot learn in one context learns readily in another — the capacity is present; the conditions that make it accessible are not consistently available.

Increased pressure or repetition alone rarely expands learning. One condition that may support it is the presence of sufficient margin to allocate to new processing: reduced concurrent demand, gradual activation slope, adequate recovery between demands, and a relational field that does not itself constitute a processing cost. Under these conditions, learning may become more accessible to the system — because a system less occupied with managing threat or absorbing overload may have more available capacity to attend to what is novel. That attention is one way curiosity can be understood.

Curiosity and Regulatory Availability

Curiosity is often interpreted as a personality trait or emotional disposition. Curiosity may be understood as one possible sign that current conditions are supporting exploratory access. Its presence does not directly measure regulatory margin, and its absence does not establish that margin is narrow.

Narrower available margin may make unpredictability more costly to absorb, while broader regulatory availability may support greater tolerance for uncertainty and exploration. Curiosity may reflect a kind of physiological availability: a system organized with sufficient margin, manageable activation slope, and enough recovery capacity to sustain exploratory engagement without requiring certainty as a precondition.

When curiosity, exploratory engagement, or spontaneous orientation toward the unfamiliar appear in a session or classroom, they may suggest that sufficient access has become available for expansion beyond immediate protective organization. When they are absent, the question shifts from motivation alone toward physiology: what is the system carrying that currently makes exploration too costly?

The Regulatory Bridge: Special Interests and the Arc Toward Expansion

When a system organizes around a domain of deep familiarity and high predictability, the regulatory conditions may shift in ways legible through the four parameters: concurrent demand may be reduced, the conditions under which a particular activation arc unfolds may allow a more gradual rise, more regulatory margin may remain available, and recovery may be better supported. The special interest is not a retreat from learning. It is the system finding a configuration under which learning becomes physiologically possible.

This is the regulatory bridge. New material introduced through the domain of the special interest may be processed under different regulatory conditions — not only as a source of motivation or reward, but as a context that may alter the conditions under which processing occurs. As the system stabilizes and margin becomes more consistently available, new material may begin to carry its own regulatory momentum. The special interest can shift from primary scaffold toward a less central support — not because it has been replaced, but because the system may no longer require it as the sole entry point.

The same logic applies to behaviors that initially appear as obstacles to engagement. A behavior emerging under conditions of regulatory depletion does not need to be extinguished. As the conditions change — as margin becomes more consistently available, as the relational field becomes more reliably safe — the behavior becomes less probable, because the conditions that made it the most available response have shifted. What replaces it is the system's natural activity when it has the resources to choose something other than protection.

What Established Approaches Illuminate — and What the Framework Explains

Several established approaches to supporting autistic individuals have arrived, through different routes, at practices consistent with what this framework predicts. Structured teaching approaches can be read as altering the conditions that contribute to concurrent demand: visual schedules, organized environments, and predictable routines may reduce uncertainty and the processing demands associated with continual prediction updating. Approaches prioritizing social communication and emotional regulation within supportive relationships can be read in relation to regulatory margin and co-regulation, insofar as greater relational predictability may change the conditions under which communication and learning occur. Naturalistic and motivation-based approaches that follow the child's lead may work similarly — reducing some sources of concurrent demand and shifting the conditions under which a given activation arc unfolds.

In each case, the framework does not replace these approaches. It offers one account of the regulatory processes that may contribute to their effectiveness — and precision at that level can support more individualized rather than purely procedural application.

Precision is not permissiveness. It is the precondition for support that actually works.

12.6 — From Framework to Practice

A practitioner who has read this far is not standing in the same place they were at the book's opening. The change is not in what they know about autism or regulation in the abstract. It is in what they can see when they are in the room.

They can see a student who will not begin a task and ask, before drawing any conclusion, what this system was carrying before the task arrived. They can see a client who appears fine and read the physiological signals that the fine is costing something. They can see a moment of withdrawal and recognize it not as disengagement but as the system doing exactly what it needs to do — and hold the field steady rather than pressing for return.

That reading does not produce automatic answers. It produces better questions — questions that extend beyond the behavioral surface to the conditions that may be shaping access and response. Every clinical encounter and every classroom contains a nervous system that has been organizing itself in relation to the conditions it has encountered. That organization may not be directly observable, but behavior always emerges within ongoing physiological organization.

What is the system carrying? What does this context require of it?

What conditions would allow it to do what it is being asked to do without that asking constituting one more cost it cannot absorb?

A practitioner who has absorbed this framework is not at the end of something. They are at the beginning of a different kind of practice — one that asks different questions, attends to different signals, and offers different kinds of support. The Clinical Workbook begins where this volume ends: with a practitioner who can formulate and revise a contextual reading of the system, and who now needs tools for working from that reading. The observation structures, practice scaffolds, and applied sequences in that volume are built directly from the four parameters developed here — not as generic techniques, but as the framework in use, adapted to patterns observed across the individual, context, and time. What this volume makes more legible, the next makes workable.

Closing — What the Reading Holds

The reading practice refers to attending to the physiological conditions from which behavior emerges, rather than to behavior in isolation.

What has become visible — in a way it could not be stated at the start — is that the framework's most consequential claim is not about autism specifically. It is about where to look. When behavior is taken as the operative ground of clinical and educational practice, the organizing conditions that produced it remain invisible. When the nervous system's regulatory dynamics are taken as the operative ground, what appeared as inconsistency becomes legible as state-dependence; what appeared as refusal becomes legible as protection. The shift in level of attention is not a refinement of existing practice. It is a reorientation of it.

Energy and cost function throughout these chapters as structural principles rather than metaphors. They account for why the same environment generates different loads in different systems — and why environmental design is not a neutral act. Every assumption embedded in a schedule, a building, a relational protocol, or an institutional culture carries physiological consequences for systems whose parameters differ from the ones the design assumed. To read the nervous system accurately is, among other things, to read environments as regulatory fields: as structures that either support or tax the bodies they contain.

The same four parameters that make physiological organization more legible can also organize questions about environmental influence. Concurrent demand directs attention to what conditions may be adding. Activation slope directs attention to how predictability and change may shape activation trajectories. Regulatory margin directs attention to the range that may remain under current conditions and accumulated history. Recovery directs attention to whether restoration appears to be available across time. Environmental adaptation therefore follows from the same reading rather than requiring a separate framework: conditions can be adjusted, their effects observed, and the reading revised accordingly.

This reorientation changes what the practitioner is responsible for considering. Not only what is visible at the behavioral surface, but the less-visible conditions that may be contributing — the starting state, the demands already accumulated, and the recovery the preceding hours may have permitted or constrained. These conditions are inferred rather than directly seen, and the reading remains open to revision as conditions are adjusted and new observations emerge.

When that level becomes the basis of observation, something follows that behavioral reading alone cannot reliably produce: disruption may become less probable as the conditions generating it change; learning may expand when margin is available to absorb it; curiosity — which may depend on supportive physiological conditions — becomes more likely to appear. Not because the person has been treated or trained differently, but because the environment has stopped constituting a cost the system cannot recover from.

Access describes the degree to which the system's current organization shapes what can be engaged with — and how — in both internal and external experience.

Safety enables recovery. Recovery restores organization. Organization expands access.

Behavior is not the thing being explained. Access is the thing being protected.

A note on what this framework does not attempt to explain: the regulatory account developed here addresses the physiological conditions under which autistic experience unfolds and the costs those conditions generate under mismatch. It does not address what autistic people care about, value, or find meaningful — the content of experience, as distinct from its organizing conditions. It does not account for the social and political dimensions of autistic life: the structural conditions that produce mismatch in the first place, the role of institutions, or the work of autistic communities and advocacy in reshaping those conditions. And it does not provide an account of autistic strengths that stands independently of the mismatch framing — it can describe what becomes possible when margin is wide and conditions are adequate, but it cannot say, within its own terms, what those possibilities are worth. These are genuine limits of a physiological reading framework, not procedural disclaimers. Other approaches — developmental, phenomenological, social, political, and those built from within autistic community — address what this one does not. The framework offered here is not a replacement for those accounts. It is one layer of a fuller understanding that none of these approaches, alone, provides.

That is where reading begins. And it is where accompaniment becomes physiologically possible.

Writing this required conditions that could be sustained; finishing it required allowing them to end.

Glossary of Core Constructs

These terms function as reading tools. Their meaning is developed throughout the book and is gathered here as a point of integration rather than introduction.

Concurrent demand

The total processing load the nervous system is managing at any moment—including sensory, cognitive, emotional, relational, and internal physiological channels — among them unresolved pain, illness, disrupted sleep, and undetected hunger. It is the baseline condition under which each new event is received and processed. Higher concurrent demand means less regulatory margin is available when additional demand arrives.

Activation slope

The rate at which the system moves from baseline toward higher activation when an incoming signal diverges from prediction. It describes how quickly threshold is approached. Momentum, by contrast, describes how long activation continues unfolding once recruited. A steeper slope means the system reaches threshold more rapidly and requires more of the available margin to absorb.

Regulatory margin

The available range within which activation can rise without exceeding what the system can sustain under its current conditions. Within the framework, margin is modeled as jointly shaped by current activation level, concurrent demand, and completeness of prior recovery.

Recovery

The extent to which the system returns toward baseline following activation.

This process may involve parasympathetic re-engagement alongside changes in autonomic tone, hormonal activity, muscular tone, respiration, and cortical arousal. Recovery varies in rate and completeness; when incomplete, it carries residue into the next cycle, elevating the baseline from which subsequent demands are met. Recovery is the process through which concurrent demand can be cleared and regulatory margin restored, to the extent that recovery is complete.

Activation

Shifts in physiological energy that prepare the organism for engagement, response, or protection, varying in intensity according to context, concurrent demand, and available regulatory margin. Activation is not binary — it moves across a continuum from baseline alertness to full protective mobilization. Its clinical significance lies not in its presence but in its rate of rise, the margin available to contain it, and the completeness with which it resolves.

Threshold

Threshold marks the point at which activation exceeds available regulatory margin and the system transitions from sustained engagement toward protective reorganization. Present conditions and accumulated physiological patterning shape threshold proximity and available margin rather than redefining the threshold itself.

Residue

The activation that remains physiologically active when recovery is incomplete — the carried-forward increment of a prior cycle that has not fully resolved. Residue is not pathology; it is the ordinary consequence of regulatory work that has not yet cleared. Within the framework, accumulated residue is modeled as elevating the baseline from which subsequent demands are met and narrowing regulatory margin, so that less additional activation may be needed for ordinary perturbations to approach or cross threshold. What appears disproportionate to the immediate event may therefore be read as one possible expression of current demand meeting the history carried forward from prior cycles.

Energetic cost

The physiological expenditure required to remain present and organized within experience. Energetic cost accumulates when recovery does not sufficiently restore what activation has required — and is often invisible precisely because it is being successfully managed. Its downstream effects are what become clinically legible: elevated baseline, narrowed margin, reduced functional access, and fatigue that follows what appeared, from the outside, to be an unremarkable day.

Energy

The available physiological capacity through which the system mobilizes, adapts, and recovers. Energy fluctuates with demand, restoration, and cumulative load — it is not a fixed resource and does not reflect motivation or emotional intensity. What depletes it is sustained activation without adequate recovery; what restores it is the completion of arcs under conditions that allow parasympathetic re-engagement alongside the broader physiological changes associated with recovery. Energy refers to capacity; energetic cost refers to expenditure.

Bodily Organization (Architecture)

Bodily organization refers to the integrated configuration through which experience becomes structured in the present, encompassing sensory registration, autonomic modulation, motor readiness, predictive adjustment, relational modulation, and temporal pacing as a single dynamically coordinated system.

Coherence

The degree of coordination among sensory, autonomic, emotional, and cognitive processes, such that experience remains inhabitable as intensity varies. Coherence is what allows the system to function across rising and falling activation without loss of internal organization. It is related to but distinct from regulation: a system can remain within available margin while coherence is nonetheless fragile if the coordinating processes are under sustained concurrent demand.

Continuity

Continuity refers to the system's capacity to carry experience forward across changing levels of activation without fragmentation. Continuity is what allows a person to remain internally connected to prior states as intensity rises and falls. It narrows under sustained concurrent demand and is most vulnerable when activation exceeds available margin repeatedly without adequate recovery.

Availability

Availability refers to the degree to which the body can remain present within experience as activation rises and falls, allowing sensing, responding, and relating to remain accessible without loss of internal continuity.

Availability is not a stable trait but a state-sensitive condition: it expands when margin is adequate and recovery is complete, and contracts as concurrent demand accumulates or threshold proximity increases.

Functional Access

Functional access refers to the availability of cognitive, emotional, and communicative processes within a given bodily state, expanding or narrowing according to how the system is organized in that moment.

Internal Time

Internal time refers to the intrinsic rhythm through which experience is processed, integrated, and completed, emerging from the system itself and reflecting both present conditions and accumulated physiological history.

Regulation

Regulation refers to the physiological capacity to move through changing states of activation — including high activation — while maintaining the organization needed to recover, restore margin, and return toward sustainable baseline.

Bodily memory

The retention of physiological organization across time — how prior patterns of activation, completion, interruption, and accompaniment continue to organize present expectation. Bodily memory does not refer to the storage of specific events but to the persistence of configurations: how the system was organized when an arc rose, whether recovery occurred, and whether accompaniment was present. These configurations remain active in baseline, threshold proximity, and anticipatory organization, shaping how present signals are registered and what the system predicts will be required. The concept is used here as a descriptive construct within the regulatory framework rather than as a reference to a discrete memory system or a claim about a specific neural mechanism.

Co-regulation

Co-regulation refers to a relational physiological process in which another person's state — through rhythm, pacing, predictability, presence, and other relational conditions — becomes part of the regulatory environment available to a system, helping sustain regulatory movement without replacing the system's own activity. This influence is bidirectional but not necessarily symmetrical: each person's state can become part of the other's regulatory environment without affecting both systems equally, in the same direction, or at the same time.

Protective Organization

Protective organization refers to the adaptive reconfiguration that emerges when the cost of remaining open exceeds available margin, prioritizing preservation of viability even as functional access temporarily narrows.

Reparative trace

Within the framework developed here, a reparative trace refers to a proposed physiological revision arising when an activation arc completes under relational conditions different from those that contributed to the current protective organization — specifically, when the cost of remaining open is lower than anticipated. Rather than implying a stored narrative, the construct describes a possible revision in the anticipatory organization of future relational encounters. With repetition, discrepancies between predicted and experienced relational cost may contribute to competing expectations about what relational presence will require. This account draws on extinction learning as an explanatory model: prior protective predictions need not be erased for new predictions, built from different relational evidence, to become available.

Closure / Integration

Closure refers to the completion of a bodily process through which activation reorganizes and returns toward relative stability, allowing the system to release what has been mobilized rather than carrying it forward as unresolved load.

Appendix A — Future Directions and Research Agenda

Inference 1: Sustained behavioral organization is associated with cumulative physiological load comparable to repeated acute stress events (§8.6). The allostatic-load literature supports the broader proposition that sustained or repeated physiological demands can contribute to cumulative multisystem burden (McEwen, 2007; Ganzel et al., 2010). This book proposes that sustained adaptive effort in high-camouflaging-demand environments may contribute to this kind of cumulative physiological cost. What would support it: longitudinal studies tracking cortisol slope, high-frequency HRV, and parasympathetic recovery rate across periods of high versus low camouflaging demand within the same autistic adults. What would revise it: evidence that camouflaging demand and physiological load are uncorrelated at the within-person level, or that the physiological signature of camouflaging effort resembles acute rather than chronic activation.

Inference 2: Autistic burnout follows a second, less-visible pathway through sustained behavioral organization that is physiologically distinct from repeated acute threshold crossings (§8.6). This account predicts a specific burnout phenotype: collapse following extended periods of behavioral organization without visible threshold crossings. What would support it: retrospective phenomenological studies distinguishing burnout onset preceded by visible dysregulation from burnout onset preceded by extended behavioral organization, with prospective ambulatory physiological tracking in autistic adults at elevated burnout risk.

Inference 3: Candidate indicators relevant to within-person regulatory state predict within-person cognitive and communicative access variability, more strongly in autistic than neurotypical samples (§§7–8). What would support it: intensive longitudinal designs (28+ daily sessions) combining morning physiological measures, self-reported recovery quality, and standardized cognitive and communicative tasks in autistic and neurotypical adults, analyzed with multilevel modeling to partition within-person and between-person variance. These measures would operationalize selected components relevant to regulatory state rather than directly measure the RNS constructs themselves.

Four Proposed Study Designs

Study 1 — The Camouflage-Load Study

Research question: Does high camouflaging demand generate sustained low-grade physiological activation measurable in chronic-load markers independent of acute stress event frequency?

Design: Ambulatory longitudinal study (n ≥ 80 autistic adults, 8-week observation periods). Camouflaging demand indexed via CAT-Q and ecological momentary assessment. Outcomes: ambulatory cortisol, continuous HRV, standardized physiological recovery rate at observation-period start and end. Within-person multilevel modeling. Critical controls: acute stress event frequency, burnout screening, perceived recovery adequacy.

Study 2 — The Burnout Pathway Study

Research question: Can autistic burnout onset types (acute-overload vs. sustained-adaptation) be reliably distinguished, and do they differ in severity, recovery trajectory, and physiological signature?

Design: Mixed-methods retrospective study (n ≥ 150 autistic adults with burnout history) with structured interviews characterizing the 12 months preceding onset. Prospective follow-up of an elevated-risk subset with monthly autonomic and behavioral assessment. Primary prediction: sustained-adaptation-pathway burnout shows greater severity and slower recovery, preceded by lower acute dysregulation and higher camouflaging demand.

Study 3 — The State-Performance Variability Study

Research question: Do candidate physiological and experiential indicators relevant to within-person regulatory state predict within-person cognitive and communicative variability more strongly in autistic than neurotypical adults?

Design: Intensive longitudinal study ($n \geq 60$ autistic adults, $n \geq 60$ matched neurotypical controls; 28 daily sessions). Morning physiological measures + self-reported recovery quality + standardized cognitive and communicative tasks. Multilevel modeling partitioning within-person and between-person variance. Primary prediction: HRV and recovery quality predict same-day performance within autistic participants more strongly than within neurotypical controls. HRV is treated here as a candidate indicator of autonomic regulation, not as a direct measure of regulatory margin or overall regulatory state.

Study 4 — The Reparative Trace Study

Research question: Does repeated exposure to relational conditions meeting the four recovery criteria produce measurable, cumulative changes in autistic adults' autonomic anticipatory organization within that specific relational context, distinguishable from general baseline improvement?

Design: Single-case and small-group longitudinal study ($n \geq 20$ autistic adults in ongoing therapeutic or clinical accompaniment relationships, 6–12 months observation period). Candidate autonomic indicators — pre-session HRV, electrodermal activity at relational approach, heart rate recovery slope within session — collected at monthly intervals across the observation period. Within-person multilevel modeling tracking trajectory of anticipatory autonomic response (measured before sessions begin, when the relational context is cued but no demand has yet been placed) against general daily baseline HRV measured on non-session days. Primary prediction: reparative trace accumulation will be context-specific — anticipatory autonomic response will improve selectively in the specific relational context before generalizing to other relational contexts, consistent with the extinction-learning account of §11.4. Secondary prediction: the rate of autonomic change will correlate with self-reported reduction in pre-session protective preparation (camouflaging effort, anticipatory fatigue) across the same interval. What would support the trace-revision account: context-specific autonomic change preceding general improvement. Pre-study CAT-Q scores could be examined as an exploratory measure of camouflaging history, but should not be treated as an index of the depth of a protective trace. What would revise it: equivalent improvement across relational contexts from the outset, suggesting general nervous system recovery rather than relationship-specific trace revision.

A Note on Collaboration

The empirical questions this framework generates will be most productively pursued through research partnerships that include autistic people as co-investigators. This work draws on published autistic accounts as primary evidence of lived organization, while recognizing that it has not been co-developed with autistic collaborators — a limitation that applies to these study designs as it does to the framework itself. Refinement by autistic researchers, including modification of outcome measures, interview designs, and interpretation frameworks, is both methodologically appropriate and scientifically necessary. This research agenda is an invitation to that collaboration.

References

Alexander, F., & French, T. M. (1946). Psychoanalytic therapy: Principles and applications. Ronald Press.

Bargiela, S., Steward, R., & Mandy, W. (2016). The experiences of late-diagnosed women with autism spectrum conditions: An investigation of the female autism phenotype. Journal of Autism and Developmental Disorders, 46(10), 3281–3294. https://doi.org/10.1007/s10803-016-2872-8

Barsalou, L. W. (2008). Grounded cognition. Annual Review of Psychology, 59, 617–645. https://doi.org/10.1146/annurev.psych.59.103006.093639

Brewer, R., Cook, R., & Bird, G. (2016). Alexithymia: A general deficit of interoception. Royal Society Open Science, 3(10), 150664. https://doi.org/10.1098/rsos.150664

Buckle, K. L., Leadbitter, K., Poliakoff, E., & Gowen, E. (2021). "No way out except from external intervention": First-hand accounts of autistic inertia. Frontiers in Psychology, 12, 631596. https://doi.org/10.3389/fpsyg.2021.631596

Clark, A. (2013). Whatever next? Predictive brains, situated agents, and the future of cognitive science. Behavioral and Brain Sciences, 36(3), 181–204. https://doi.org/10.1017/s0140525x12000477

Coan, J. A., & Sbarra, D. A. (2015). Social baseline theory: The social regulation of risk and effort. Current Opinion in Psychology, 1, 87–91. https://doi.org/10.1016/j.copsyc.2014.12.021

Craig, A. D. (2009). How do you feel — now? The anterior insula and human awareness. Nature Reviews Neuroscience, 10(1), 59–70. https://doi.org/10.1038/nrn2555

Crompton, C. J., Hallett, S., McAuliffe, C., Stanfield, A. C., & Fletcher-Watson, S. (2022). "A group of fellow travellers who understand": Interviews with autistic people about post-diagnostic peer support in adulthood. Frontiers in Psychology, 13, 831628. https://doi.org/10.3389/fpsyg.2022.831628

Damasio, A. (2010). Self comes to mind: Constructing the conscious brain. Pantheon Books.

Feldman, R. (2007). Parent–infant synchrony: Biological foundations and developmental outcomes. Current Directions in Psychological Science, 16(6), 340–345. https://doi.org/10.1111/j.1467-8721.2007.00532.x

Feldman, R., Magori-Cohen, R., Galili, G., Singer, M., & Louzoun, Y. (2011). Mother and infant coordinate heart rhythms through episodes of interaction synchrony. Infant Behavior and Development, 34(4), 569–577. https://doi.org/10.1016/j.infbeh.2011.06.008

Friston, K. (2010). The free-energy principle: A unified brain theory? Nature Reviews Neuroscience, 11(2), 127–138. https://doi.org/10.1038/nrn2787

Fuchs, T. (2012). The phenomenology of body memory. In S. C. Koch, T. Fuchs, M. Summa, & C. Müller (Eds.), Body memory, metaphor and movement (pp. 9–22). John Benjamins.

Ganzel, B. L., Morris, P. A., & Wethington, E. (2010). Allostasis and the human brain: Integrating models of stress from the social and life sciences. Psychological Review, 117(1), 134–174. https://doi.org/10.1037/a0017773

Gonçalves, J. S., Moura, P. J., & Moioli, M. (2023). Somatosensory and sensory processing differences in autism spectrum disorder: A systematic review. Journal of Autism and Developmental Disorders, 53(4), 1395–1418. https://doi.org/10.1007/s10803-022-05624-8

Grandin, T. (1995). Thinking in pictures: And other reports from my life with autism. Doubleday.

Grillon, C. (2008). Models and mechanisms of anxiety: Evidence from startle studies. Psychopharmacology, 199(3), 421–437. https://doi.org/10.1007/s00213-007-1019-1

Grossman, P., & Taylor, E. W. (2007). Toward understanding respiratory sinus arrhythmia: Relations to cardiac vagal tone, evolution and biobehavioral functions. Biological Psychology, 74(2), 263–285. https://doi.org/10.1016/j.biopsycho.2005.11.014

Hadjikhani, N., Åsberg Johnels, J., Zürcher, N. R., Lassalle, A., Guillon, Q., Hippolyte, L., Billstedt, E., Ward, N., Lemonnier, E., & Gillberg, C. (2017). Look me in the eyes: Constraining gaze in the eye-region provokes abnormally high subcortical activation in autism. Scientific Reports, 7, 3163. https://doi.org/10.1038/s41598-017-03378-5

Higgins, J. M., Arnold, S. R. C., Weise, J., Pellicano, E., & Trollor, J. N. (2021). Defining autistic burnout through experts by lived experience: Grounded Delphi method investigating #AutisticBurnout. Autism, 25(8), 2356–2369. https://doi.org/10.1177/13623613211019858

Hull, L., Petrides, K. V., Allison, C., Smith, P., Baron-Cohen, S., Lai, M.-C., & Mandy, W. (2017). Putting on my best normal: Social camouflaging in adults with autism spectrum conditions. Journal of Autism and

Developmental Disorders, 47(8), 2519–2534. https://doi.org/10.1007/s10803-017-3166-5

Hull, L., Lai, M.-C., Baron-Cohen, S., Allison, C., Smith, P., Petrides, K. V., & Mandy, W. (2020). Gender differences in self-reported camouflaging in autistic and non-autistic adults. Autism, 24(2), 352–363. https://doi.org/10.1177/1362361319864804

Karalunas, S. L., Geurts, H. M., Konrad, K., Bender, S., & Nigg, J. T. (2014). Annual research review: Reaction time variability in ADHD and autism spectrum disorders — measurement and mechanisms of a proposed trans-diagnostic phenotype. Journal of Child Psychology and Psychiatry, 55(6), 685–710. https://doi.org/10.1111/jcpp.12217

Lai, M.-C., Lombardo, M. V., Ruigrok, A. N. V., Chakrabarti, B., Auyeung, B., Szatmari, P., Happé, F., Baron-Cohen, S., & MRC AIMS Consortium. (2017). Quantifying and exploring camouflaging in men and women with autism. Autism, 21(6), 690–702. https://doi.org/10.1177/1362361316671012

Lawson, R. P., Rees, G., & Friston, K. J. (2014). An aberrant precision account of autism. Frontiers in Human Neuroscience, 8, 302. https://doi.org/10.3389/fnhum.2014.00302

Linden, W., Earle, T. L., Gerin, W., & Christenfeld, N. (1997). Physiological stress reactivity and recovery: Conceptual siblings separated at birth? Journal of Psychosomatic Research, 42(2), 117–135. https://doi.org/10.1016/S0022-3999(96)00240-1

Lisi, E. (2020, February). What it's really like to have autism [Video]. TED Conferences.

https://www.ted.com/talks/ethan_lisi_what_it_s_really_like_to_have_autism

Mantzalas, J., Richdale, A. L., Adikari, A., Lowe, J., & Dissanayake, C. (2022). What is autistic burnout? A thematic analysis of posts on two online platforms. Autism in Adulthood, 4(1), 52–65. https://doi.org/10.1089/aut.2021.0021

Mazefsky, C. A., Herrington, J., Siegel, M., Scarpa, A., Maddox, B. B., Scahill, L., & White, S. W. (2013). The role of emotion regulation in autism spectrum disorder. Journal of the American Academy of Child & Adolescent Psychiatry, 52(7), 679–688. https://doi.org/10.1016/j.jaac.2013.05.006

McEwen, B. S. (2007). Physiology and neurobiology of stress and adaptation: Central role of the brain. Physiological Reviews, 87(3), 873–904. https://doi.org/10.1152/physrev.00041.2006

McEwen, B. S., & Stellar, E. (1993). Stress and the individual: Mechanisms leading to disease. Archives of Internal Medicine, 153(18), 2093–2101. https://doi.org/10.1001/archinte.1993.00410180039004

Merchie, A., Heremans, E., Wenderoth, N., & Smeets, J. B. J. (2023). Sensory processing and habituation in autism spectrum disorder: A meta-analysis. Psychological Bulletin, 149(5–6), 259–310. https://doi.org/10.1037/bul0000394

Milton, D. E. M. (2012). On the ontological status of autism: The 'double empathy problem'. Disability & Society, 27(6), 883–887. https://doi.org/10.1080/09687599.2012.710008

Mineka, S., & Kihlstrom, J. F. (1978). Unpredictable and uncontrollable events: A new perspective on experimental neurosis. Journal of Abnormal Psychology, 87(2), 256–271. https://doi.org/10.1037/0021-843X.87.2.256

Mottron, L., Dawson, M., Soulières, I., Hubert, B., & Burack, J. (2006). Enhanced perceptual functioning in autism: An update, and eight principles of autistic perception. Journal of Autism and Developmental Disorders, 36(1), 27–43. https://doi.org/10.1007/s10803-005-0040-7

Murray, D., Lesser, M., & Lawson, W. (2005). Attention, monotropism and the diagnostic criteria for autism. Autism, 9(2), 139–156. https://doi.org/10.1177/1362361305051398

Pearson, A., & Rose, K. (2021). A conceptual analysis of autistic masking: Understanding the narrative of stigma and the illusion of choice. Autism in Adulthood, 3(1), 52–60. https://doi.org/10.1089/aut.2020.0043

Pellicano, E., & Burr, D. (2012). When the world becomes 'too real': A Bayesian explanation of autistic perception. Trends in Cognitive Sciences, 16(10), 504–510. https://doi.org/10.1016/j.tics.2012.08.009

Porges, S. W. (2011). The polyvagal theory: Neurophysiological foundations of emotions, attachment, communication, and self-regulation. W. W. Norton.

Prizant, B. M., Wetherby, A. M., Rubin, E., Laurent, A. C., & Rydell, P. J. (2006). The SCERTS model: A comprehensive educational approach for children with autism spectrum disorders. Paul H. Brookes.

Ramsay, D. S., & Woods, S. C. (2014). Clarifying the roles of homeostasis and allostasis in physiological regulation. Psychological Review, 121(2), 225–

247. https://doi.org/10.1037/a0035942

Rapaport, H., Clapham, H., Adams, J., Lawson, W., Porayska-Pomsta, K., & Pellicano, E. (2024). "I live in extremes": A qualitative investigation of Autistic adults' experiences of inertial rest and motion. Autism, 28(5), 1305–1315. https://doi.org/10.1177/13623613231198916

Raymaker, D. M., Teo, A. R., Steckler, N. A., Lentz, B., Scharer, M., Delos Santos, A., Kapp, S. K., Hunter, M., Joyce, A., & Nicolaidis, C. (2020). "Having all of your internal resources exhausted beyond measure and being left with no clean-up crew": Defining autistic burnout. Autism in Adulthood, 2(2), 132–143. https://doi.org/10.1089/aut.2019.0079

Sapolsky, R. M. (2004). Why zebras don't get ulcers: The acclaimed guide to stress, stress-related diseases, and coping (3rd ed.). Henry Holt.

Schore, A. N. (2012). The science of the art of psychotherapy. W. W. Norton.

Seth, A. K. (2013). Interoceptive inference, emotion, and the embodied self. Trends in Cognitive Sciences, 17(11), 565–573. https://doi.org/10.1016/j.tics.2013.09.007

Seth, A. K. (2021). Being you: A new science of consciousness. Dutton.

Seth, A. K., & Friston, K. J. (2016). Active interoceptive inference and the emotional brain. Philosophical Transactions of the Royal Society B, 371(1708), 20160007. https://doi.org/10.1098/rstb.2016.0007

Seth, A. K., Suzuki, K., & Critchley, H. D. (2012). An interoceptive predictive coding model of conscious presence. Frontiers in Psychology, 2, 395. https://doi.org/10.3389/fpsyg.2011.00395

Shaffer, F., & Ginsberg, J. P. (2017). An overview of heart rate variability metrics and norms. Frontiers in Public Health, 5, 258. https://doi.org/10.3389/fpubh.2017.00258

Siegel, D. J. (2012). The developing mind: How relationships and the brain interact to shape who we are (2nd ed.). Guilford Press.

Sterling, P. (2012). Allostasis: A model of predictive regulation. Physiology & Behavior, 106(1), 5–15. https://doi.org/10.1016/j.physbeh.2011.06.004

Stevenson, R. A., Segers, M., Ferber, S., Barense, M. D., Camarata, S., & Wallace, M. T. (2016). Keeping time in the brain: Autism spectrum disorder and audiovisual temporal processing. Autism Research, 9(7), 720–738. https://doi.org/10.1002/aur.1566

Thayer, J. F., & Lane, R. D. (2000). A model of neurovisceral integration in emotion regulation and dysregulation. Journal of Affective Disorders, 61(3), 201–216. https://doi.org/10.1016/s0165-0327(00)00338-4

Thompson, R. F., & Spencer, W. A. (1966). Habituation: A model phenomenon for the study of neuronal substrates of behavior. Psychological Review, 73(1), 16–43. https://doi.org/10.1037/h0022681

Thompson, E. (2007). Mind in life: Biology, phenomenology, and the sciences of mind. Harvard University Press.

Varela, F. J., Thompson, E., & Rosch, E. (1991). The embodied mind: Cognitive science and human experience. MIT Press.

Williams, D. (1992). Nobody nowhere: The extraordinary autobiography of an autistic. Doubleday.

Yergeau, M. R. (2018). Authoring autism: On rhetoric and neurological queerness. Duke University Press.

Zillmann, D. (1971). Excitation transfer in communication-mediated aggressive behavior. Journal of Experimental Social Psychology, 7(4), 419–434. https://doi.org/10.1016/0022-1031(71)90075-8

About the Author

María E. Reyes, M.S., CCC-SLP, is a bilingual speech-language pathologist with clinical experience across healthcare, educational, and community settings. She immigrated to the United States, where she developed her academic and professional career. As she learned English as a second language, she became increasingly attuned to the role of bodily signals in communication—an experience that continues to inform her clinical work.

Her practice focuses on nervous system regulation, embodied communication, and the conditions that support functional access in both children and adults. She has worked extensively with autistic individuals and their families and has provided training and consultation to clinical and educational teams.

This book is intended as a conceptual and observational tool for professionals and graduate-level students, and for readers seeking a deeper understanding of how experience is organized through the body. It reflects years of clinical observation and a sustained inquiry into how nervous system processes organize experience across time, and how those processes can be read in real-world contexts.

She may be contacted through the author's website at ReadingTheNervousSystem.com for inquiries related to training, consultation, and the companion learning ecosystem described in this volume.

The Reading the Nervous System Series

Volume I — Reading the Nervous System: A Physiological Reading of Autistic Experience (this volume). The theoretical foundation. Introduces the four-parameter regulatory framework — concurrent demand, activation slope, regulatory margin, and recovery — and develops its application to autistic nervous system organization across twelve chapters. Intended to be read in full before Volume II.

Volume II — Reading the Nervous System: The Clinical Workbook The applied companion to Volume I. Chapter 12 of this volume traced what becomes possible when nervous system organization is read with greater contextual precision — disruption may become less probable as the conditions generating it change, learning may expand when margin is available to absorb it, curiosity may become more available as regulatory conditions support exploration, and special interests may become regulatory bridges rather than obstacles to engagement. The Clinical Workbook is where that possibility becomes practice: structured observations, environment design tools, reflective exercises, and session-level guidance built directly from the four parameters developed here — not as generic techniques, but as the framework in use, adapted to observed patterns across the individual, context, and time. What this volume made legible, that one makes workable. Assumes fluency with Volume I. Designed for practitioners in speech-language pathology, occupational therapy, educational psychology, and related fields who accompany autistic children and adults.

The two volumes are designed to be used in sequence. Volume I provides the reading. Volume II provides the practice. Together, they constitute a complete clinical learning sequence for regulation-informed accompaniment.

The Full Learning Ecosystem

The Reading the Nervous System framework is supported by a multi-component learning ecosystem designed for practitioners at every stage of engagement with the material.

Video lecture series. A structured video series aligned with each Part of Volume I, providing conceptual commentary, clinical examples, and guided reflection for practitioners working through the theoretical framework. Designed for individual self-study or group learning contexts.

Clinical courses. Structured continuing education courses for practitioners in speech-language pathology, occupational therapy, and education, applying the framework to specific clinical contexts including assessment, session design, environment consultation, and practitioner self-regulation.

The recommended sequence for full engagement with the ecosystem is: Volume I → Video lecture series → The Clinical Workbook → Clinical courses. Practitioners who encounter The Clinical Workbook before Volume I are strongly encouraged to return to the theoretical foundation before applying the workbook's tools, as the exercises depend on a level of framework fluency that the workbook itself does not build from scratch. Practitioners who have completed Volume I will find that Chapter 12 — What Legibility Opens — provides the conceptual bridge between the theoretical framework and its clinical application, and is recommended as a direct entry point into The Clinical Workbook.

For current availability, course schedules, and more, visit ReadingTheNervousSystem.com.

READING THE NERVOUS SYSTEM

ReadingTheNervousSystem.com

Continue exploring the framework.

Enlarged versions of selected figures, companion learning resources, and materials designed to support the continued exploration and application of the framework are available at:

ReadingTheNervousSystem.com

For professional education, clinical trainings, and workshops related to Reading the Nervous System, visit:

ReadingTheNervousSystem.com

Published by Neurodiversidad Consciente